Harold Broberg, PE
Indiana-Purdue University
at Fort Wayne

EXPERIMENTS IN ELECTRONIC COMMUNICATION

MERRILL PUBLISHING COMPANY
Columbus   Toronto   London   Melbourne

Published by Merrill Publishing Company
Columbus, Ohio 43216

This book was set in Times Roman

Administrative Editor: David Garza
Production Editor: Mary M. Irvin
Art Coordinator: Vincent A. Smith
Cover Designer: Brian Deep

Library of Congress Catalog Card Number:
International Standard Book Number: 0-675-89-64069
Printed in the United States of America
1 2 3 4 5 6 7 8 9—94 93 92 91 90

MERRILL'S INTERNATIONAL SERIES
IN ELECTRICAL AND ELECTRONICS TECHNOLOGY

ADAMSON	*Applied Pascal for Technology,* 20771–1 *Structured BASIC Applied to Technology,* 20772–X *Structured C for Technology,* 20993-5 *Structured C for Technology* (w/disks), 21289–8
ANTONAKOS	*The 68000 Microprocessor: Hardware and Software, Principles and Applications,* 21043-7
ASSER/STIGLIANO/ BAHRENBURG	*Microcomputer Servicing: Practical Systems and Troubleshooting,* 20907–2 *Microcomputer Theory and Servicing,* 20659–6 *Lab Manual to accompany Microcomputer Theory and Servicing,* 21109–3
ASTON	*Principles of Biomedical Instrumentation and Measurement,* 20943–9
BATESON	*Introduction to Control System Technology, Third Edition,* 21010–0
BEACH/JUSTICE	*DC/AC Circuit Essentials,* 20193–4
BERLIN	*Experiments in Electronic Devices, Second Edition,* 20881–5 *The Illustrated Electronics Dictionary,* 20451–8
BERLIN/GETZ	*Experiments in Instrumentation and Measurement,* 20450–X *Fundamentals of Operational Amplifiers and Linear Integrated Circuits,* 21002–X *Principles of Electronic Instrumentation and Measurement,* 20449–6
BOGART	*Electronic Devices and Circuits, Second Edition,* 21150–6
BOGART/BROWN	*Experiments in Electronic Devices and Circuits, Second Edition,* 21151–4
BOYLESTAD	*DC/AC: The Basics,* 20918–8 *Introductory Circuit Analysis, Sixth Edition,* 21181–6
BOYLESTAD/ KOUSOUROU	*Experiments in Circuit Analysis, Sixth Edition,* 21182–4 *Experiments in DC/AC Basics,* 21131–X
BREY	*8086/8088 Microprocessor: Architecture, Programming, and Interfacing,* 20443–7 *Microprocessors and Peripherals: Hardware, Software, Interfacing, and Applications, Second Edition,* 20884–X
BROBERG	*Experiments in Electronic Communication,* 21257–X
BUCHLA	*Digital Experiments: Emphasizing Systems and Design, Second Edition,* 21180–8 *Experiments in Electric Circuits Fundamentals,* 20836–X *Experiments in Electronics Fundamentals: Circuits, Devices and Applications,* 20736–3
COX	*Digital Experiments: Emphasizing Troubleshooting, Second Edition,* 21196–4
DELKER	*Experiments in 8085 Microprocessor Programming and Interfacing,* 20663–4

PREFACE

The experiments in this manual are designed to provide a solid foundation in electronic circuits and techniques used in communications. The emphasis is on the fundamentals and on providing labs that have been well tested so that problems in circuitry and measurement will be provided by students and not the manual. Procedures are detailed, but these are not cookbook laboratories, and the students will need a good foundation of theory and understanding to complete them properly. There are more than enough experiments included for two semesters of challenging laboratory work that will provide insight and hands-on knowledge of theory, techniques, and some modern integrated circuits. Experiments on antenna radiation patterns, transmission line/waveguide, television, and programming were not included because most college programs have specific equipment for use in these areas.

Interactive PSPICE* (Evaluation Version) was used to produce many of the graphs. Because of its resolution and ease of use, it has proven to be a valuable tool in helping students learn and visualize the fundamentals.

OrCAD** was used for the electronic drawings and is an excellent schematic design tool.

Specification sheets for all the major devices used are included in the Appendix.

It is assumed throughout the experiments that students have a dual-channel oscilloscope, a frequency counter, power supplies, and a device to measure inductance and capacitance. It is also assumed that students will use short leads and good wiring practices on the higher frequency experiments. All experiments have been tested and work well using standard breadboards that can be purchased at local electronics stores or by mail order.

---

* Interactive PSPICE (Evaluation Version), 1989, MicroSim Corporation, 20 Fairbanks, Irvine, CA 92718.
** OrCAD Systems Corporation, 1987, 1049 S.W. Baseline Street, Suite 500, Hillsboro, OR 97123.

It is also assumed that the format for the lab report will be specified by the instructor. This manual provides a complete guide for performing the experiment with some explanation of circuit operation. A lab report should include problems found in circuit construction and solutions. There will be plenty for the student to report, even with the detailed procedures included here.

The approximate number of 3-hour lab periods for planning a semester or quarter schedule are as follows.

Experiment	Number of 3-Hour Periods	Number of Experiment	3-Hour Periods
1	2	12	1
2	1	13	2
3	2	14	3
4	1	15	3
5	1	16	1
6	2	17	1
7	1	18	1
8	1	19	1
9	1	20	2
10	1	21	1
11	2		

The order of the labs generally follows the text *Electronic Communication Techniques* by Paul Young. Many of these labs have been used with other texts, and general references to some texts are included with each experiment. The texts referred to in each experiment are keyed to the references given.

I greatly appreciate the efforts of all our fine students in completing earlier versions of these experiments, the assistance and suggestions of Casey Chesney in performing the final test and evaluation, and the support of my wife, Hope, during the many hours devoted to this manual.

References

1. Paul Young, *Electronic Communication Techniques,* 2d ed. (Columbus, Ohio: Merrill, 1990).
2. Dennis Roddy and John Coolen, *Electronic Communications,* 3d ed. (Reston, Va.: Reston, 1984).
3. Gary Miller, *Modern Electronic Communications,* 3d ed. (Englewood Cliffs, N.J.: Prentice Hall, 1988).
4. Wayne Tomasi, *Fundamentals of Electronic Communications Systems.* (Englewood Cliffs, N.J.: Prentice Hall, 1988). Wayne Tomasi, *Advanced Electronic Communications Systems.* (Englewood Cliffs, N.J.: Prentice Hall, 1987).
5. Harold Killen, *Modern Electronic Communication Techniques.* (New York: Macmillan, 1985).
6. Thomas Adamson, *Electronic Communications.* (Albany, N.Y.: Delmar, 1988).

CONTENTS

1

FREQUENCY RESPONSE
AND FILTERS

INTRODUCTION

The frequency domain is used extensively in the field of electronic communications. You were introduced to frequency response and filters in ac circuit analysis courses, but you may not be completely familiar with the frequency domain. The purpose of this experiment is to refresh your memory on the frequency response of common types of passive filters, to introduce you to some standard terminology used in analysis and design of filters, and to prepare you to think in the frequency domain. You will also be introduced to frequency and impedance scaling, an important subject in filter design.

After completing this experiment, you will be able to

1. Predict the general frequency response of a variety of passive filters.
2. Analyze the graphical frequency response of filters to determine (as applicable)
   a. 3-dB/cutoff/half-power frequencies
   b. band-pass frequencies
   c. band-stop frequencies
   d. order of a filter
   e. slope of a filter in decibels/decade or decibels/octave
3. Scale a circuit to perform at a specified frequency.
4. Scale a circuit to obtain specified circuit impedances.
5. Use a normalized circuit to design a filter to operate at a desired frequency and contain specified impedances.

REFERENCES

1. Young, Chapter 1.
2. Roddy and Coolen, Chapter 1.
3. Miller, Chapter 1.
4. Tomasi (*Fundamentals*), Chapter 1.
5. Killen, Chapters 2 and 3.
6. Adamson, Chapters 1 and 16.
7. Fink and Christiansen, *Electronic Engineer's Handbook,* McGraw-Hill, 1982.

MATERIALS OR SPECIAL INSTRUMENTATION

Resistors: two 5k potentiometers, one 10 Ω

Capacitors: two 1 μF, one 0.001 μF

Inductors: one 10 mH

THEORETICAL BACKGROUND

Plotting Frequency Response

The circuits discussed and used in this experiment are shown in Figure 1–1. Frequency-response graphs plot the amplitude (magnitude) of the voltage or current on the vertical axis versus the frequency on the horizontal axis, as shown in Figure 1–2. This graph shows the frequency response of the 1-pole low-pass filter circuit of Figure 1–1. Note

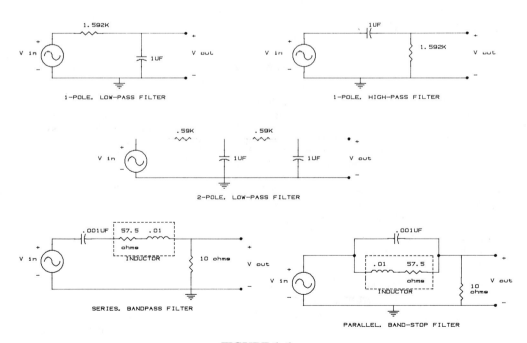

FIGURE 1–1

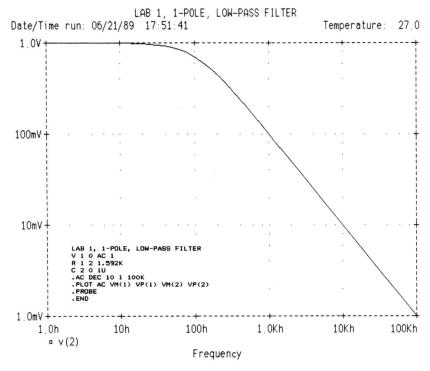

LAB 1, 1-POLE, LOW-PASS FILTER
V 1 0 AC 1
R 1 2 1.592K
C 2 0 1U
.AC DEC 10 1 100K
.PLOT AC VM(1) VP(1) VM(2) VP(2)
.PROBE
.END

□ v(2)

Frequency

FIGURE 1–2

that the vertical axis is a logarithmic scale. The amplitude can also be given in decibels, as defined in Equation 1–1 ($P = V^2/R = I^2R$; therefore, P is proportional to V^2).

$$dB = 10 \log_{10} \left(\frac{P_0}{P_1}\right) \quad \text{or} \quad dB = 10 \log_{10} \left(\frac{V_0^2}{V_1^2}\right) = 20 \log_{10} \left(\frac{V_0}{V_1}\right) \qquad (1-1)$$

For example, the vertical axis could be labeled 1 V = 0 dB, 0.1 V = −20 dB, or 0.01 V = −40 dB. The horizontal (frequency) axis can be labeled in hertz (as shown) or in radians per second. This axis is also a logarithmic scale. Semilogarithmic paper is generally used for plotting frequency-response graphs, with decibels on the vertical axis.

Filters

The commonly used filter-response terms *low-pass* (LP), *high-pass* (HP), *band-pass* (BP), and *band-stop* (BS) can be thought of as verbal descriptions of the graph of amplitude versus frequency. For instance, the graph shown in Figure 1–2 is called a low-pass filter because at low frequencies (relative to the portion of the curve that begins to decrease), the output amplitude is high (which means that the input frequencies are being "passed" to the output with little or no loss of amplitude). At high frequencies (relative to the point at which the curve begins to decrease), the output amplitude is much less than the input. This means that any high frequency is attenuated by the filter and is not passed through to the output.

3

Another term that is commonly used with filters and frequency response plots is the *cutoff, 3 dB*, or *half-power point*. These three expressions all refer to the point on the amplitude curve where the amplitude has decreased by 3 dB from the maximum voltage or current. This 3 dB drop represents 0.707 of the maximum value [20 log(0.707) = 3], which is one-half of the maximum power (since power is proportional to the voltage or current squared), as shown in Equation 1–2.

$$20 \log(0.707) = 3 = 10 \log(0.707)^2 \qquad \textbf{(1–2)}$$

The phase response of the same 1-pole low-pass filter is shown in Figure 1–3. At low frequencies, there is no phase shift. At the 3 dB frequency, there is a 45° phase shift, and at high frequencies, there is a 90° phase shift. The phase shift introduced in a circuit is very important in analysis and design.

The 1-pole high-pass filter shown in Figure 1–1 has the response plotted in Figure 1–4. This shows that at frequencies much higher than the 3 dB frequency, the output is a maximum, and at frequencies much lower than the 3 dB frequency, the output amplitude is much less than the input. Figure 1–5 shows the phase response of this high-pass filter.

Another factor of great importance in filters is how fast the curve drops off after the cutoff (3 dB) frequency. This is usually expressed in decibels per decade or decibels per octave. Decade refers to a factor of 10 in frequency, and octave refers to a factor of 2 in frequency. On a frequency-response plot, this is the slope of the curve. In both Figures 1–2 and 1–4 the slope approaches 20 dB/decade, or 6 dB/octave, as you get

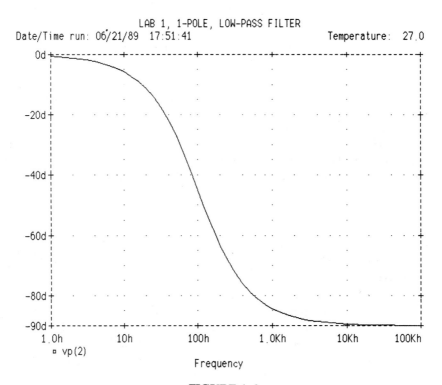

FIGURE 1–3

4

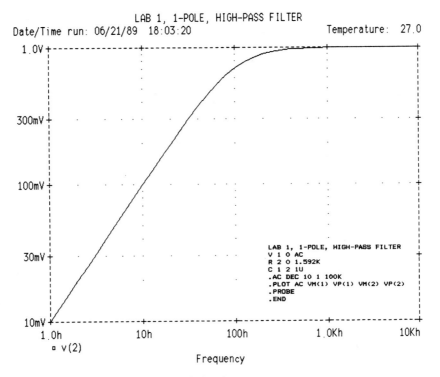

FIGURE 1–4

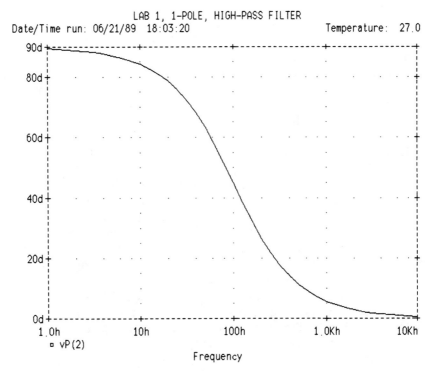

FIGURE 1–5

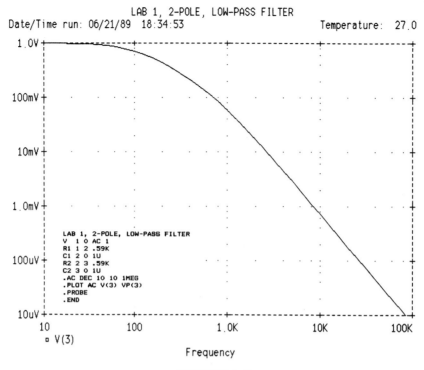

LAB 1, 2-POLE, LOW-PASS FILTER

FIGURE 1–6

farther from the 3 dB point (higher frequencies in the LP filter and lower frequencies in the HP filter). Note that the sign of the slope is different for HP and LP filters.

A filter having a slope of 20 dB/decade, or 6 dB/octave, is referred to as a *single-pole* (or *first-order*) *filter*. The "pole" designation refers to Laplace transform analysis, and the first-order terminology refers to a differential equation in which the highest derivative is a first derivative. Laplace transforms and differential equations are not essential to this discussion; however, the common usage of these terms makes it important to understand why they are used. Graphs of the magnitude and phase response of the 2-pole (or second-order) low-pass filter shown in Figure 1–1 are shown in Figures 1–6 and 1–7.

Note that in a second-order filter there are two "reactive" (capacitor or inductor) elements. This provides an indication that a circuit is a 2-pole (or second-order) filter. Analysis of the slope of Figure 1–6 shows that it approaches 40 dB/decade, or 12 dB/octave, at higher frequencies, which is double that of a single-pole filter. Similarly, a 3-pole filter would have a slope three times that of a 1-pole filter. The importance of higher-order filters is their ability to provide much greater attenuation of undesired frequencies present at the input to the filter.

For instance, Figures 1–2 and 1–6 have identical 3 dB frequencies of 100 Hz; however, an undesired input frequency of 1 kHz would be attenuated from 1 V to about 100 mV by the single-pole filter and to about 60 mV by the 2-pole filter. Similarly, a 10 kHz signal would be attenuated to about 10 mV by the first-order filter and to less than 1 mV by the 2-pole filter.

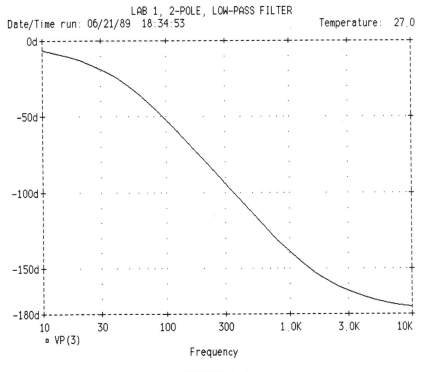

FIGURE 1–7

A graph of the series band-pass filter in Figure 1–1 is plotted in Figure 1–8. Band-pass filters "pass" a certain range of frequencies through from the input to the output, but the amplitude is attenuated (drops off) at both higher and lower frequencies. The *pass-band* of a BP filter is most often defined as the frequencies between the upper and lower 3 dB frequencies. The number of poles or order of a BP filter refers to the slope in the same way as with the HP and LP filters, except the BP filter slope falls off on both sides of a center frequency. The phase response of the BP filter is shown in Figure 1–9.

A plot of the band-stop filter shown in Figure 1–1 is shown in Figure 1–10. Band-stop filters allow all frequencies to pass through from the input to the output except a certain band that is attenuated. An example of band-stop filter use is to filter out 60-Hz hum from a stereo system. This band of frequencies is defined to be between the upper and lower 3 dB points, which are the frequencies at which the amplitude is within 3 dB of the maximum. The phase response of the band-stop filter is shown in Figure 1–11.

Scaling

Frequency Scaling

A filter often has a desired characteristic such as number of poles (order), but the frequency of the 3 dB point (or points) is not at the correct location. Rather than designing a new filter from scratch, you can use frequency scaling to convert the filter to operate in the desired frequency range. To convert any filter to have identical frequency

7

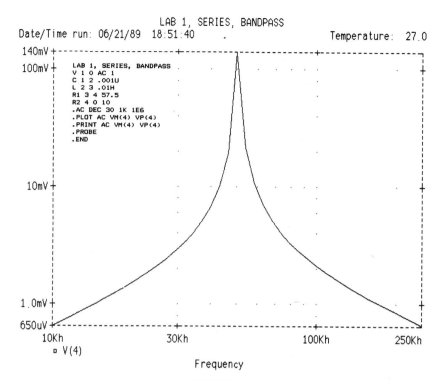

FIGURE 1–8

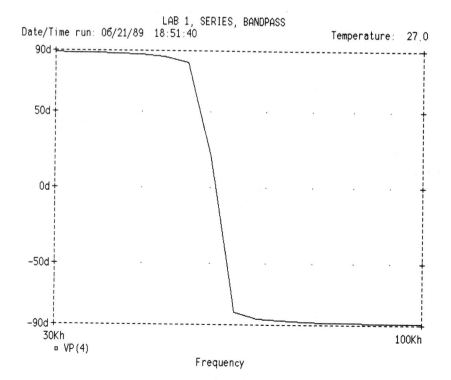

FIGURE 1–9

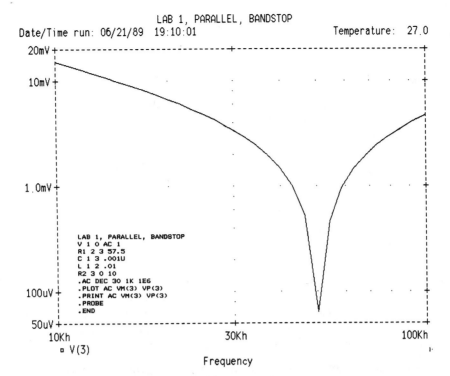

FIGURE 1–10

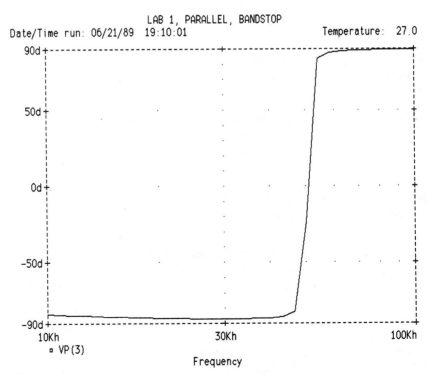

FIGURE 1–11

characteristics at a multiple M (this can be greater or less than 1) of the original frequency, perform the following steps:

1. Find the factor $M = \dfrac{\text{desired frequency}}{\text{original frequency}}$.
2. Multiply all capacitances in the original filter by $1/M$.
3. Multiply all inductances in the original filter by $1/M$.
4. Do not change resistor values.

Example 1: A 1-pole low-pass filter consists of a 1-kΩ resistor and a 1 μF capacitor with a known 3 dB frequency of 159.2 Hz. You need a 1-pole low-pass filter with a 3 dB frequency of 100 kHz. Find the R and C values.

a. $M = 10^5/159.2 = 628.1$
b. Multiply the single capacitance by $1/628.3 = 1.592 \times 10^{-3}$. This yields a new capacitance of 1.592 nF, and you have the new filter.

Impedance Scaling

At times, frequency scaling produces capacitor and inductor values that are not standard, or the circuit impedance must be increased or decreased to enable matching of the input or output circuits. To revise the impedances of a circuit while *not* changing the frequency response of the circuit, perform the following steps:

1. Determine the multiple N (greater or less than 1) by which you wish to change the overall impedance of the circuit.
2. Multiply all resistances in the circuit by N.
3. Multiply all inductances in the circuit by N.
4. Divide all capacitances in the circuit by N.

Example 2: You need the 100 kHz circuit of Example 1 to use a standard 0.001 μF capacitor, but the resistance value can use a potentiometer, so the resistance is adjustable.

1. Since you are reducing the capacitance from 1.592 nF to 1 nF, you are *increasing* the impedance of the circuit by a factor of N:

$$N = \frac{1.592 \times 10^{-9}}{10^{-9}} = 1.592$$

2. Multiplying the resistance yields 1.592 kΩ.
3. Dividing the capacitance yields the desired 0.001 μF.

If you check the 3 dB frequency of this filter, you will find that it is still 100 kHz. Impedance scaling is commonly used to change the input impedance of a circuit while retaining its operating frequency range. In this way a circuit's output impedance can be matched to the complex conjugate of the input impedance of the filter for maximum power transfer.

Normalized Filters

Filter handbooks frequently provide normalized filters (filters with component values calculated so that the cutoff frequency is 1 rad/s) as references. If you need a different frequency and impedance for your actual filter, you can use frequency and impedance scaling to create a filter with the desired specifications.

PROCEDURE AND QUESTIONS/PROBLEMS

Procedure

1. Measure the resistance, capacitance, and inductance of all components and complete the data table.

Theoretical Value	Actual Value
10 Ω	
1 μF (C_1)	
1 μF (C_2)	
0.001 μF	
10 mH/57.5 Ω	

(NOTE: *For each measurement taken,* use a DMM to ensure that the input voltage is 1 V (rms), with zero dc component, and use a frequency counter to check that the frequency is correct.

2. Construct the 1-pole low-pass filter shown in Figure 1–1 and verify the frequency and phase responses shown in Figures 1–2 and 1–3 at the four given frequencies.

Frequency	Gain/Phase from Graph	Measured Gain/Phase
10 Hz		
100 Hz		
1 kHz		
10 kHz		

Explain any differences.

3. Construct the 1-pole high-pass filter shown in Figure 1–1 and verify the frequency and phase responses shown in Figures 1–4 and 1–5 at the four given frequencies.

Frequency	Gain/Phase from Graph	Measured Gain/Phase
10 Hz		
100 Hz		
1 kHz		
10 kHz		

Explain any differences.

4. Construct the 2-pole low-pass filter shown in Figure 1–1 and verify the frequency and phase responses shown in Figures 1–6 and 1–7 at the four given frequencies.

Frequency	Gain/Phase from Graph	Measured Gain/Phase
10 Hz		
100 Hz		
1 kHz		
10 kHz		

Explain any differences.

11

5. Construct the series band-pass filter shown in Figure 1–1 and verify the frequency and phase responses shown in Figures 1–8 and 1–9 at the four given frequencies.

Frequency	Gain/Phase from Graph	Measured Gain/Phase
10 Hz		
100 Hz		
1 kHz		
10 kHz		

Explain any differences.

6. Construct the parallel band-stop filter shown in Figure 1–1 and verify the frequency and phase responses shown in Figures 1–10 and 1–11 at the four given frequencies.

Frequency	Gain/Phase from Graph	Measured Gain/Phase
10 Hz		
100 Hz		
1 kHz		
10 kHz		

Explain any differences.

Questions/Problems

1. Scale the 1-pole high-pass filter in Figure 1–1 to have a 3 dB frequency of 100 kHz (the 3 dB frequency of Figure 1–1 is 100 Hz). Use a 0.001 μF capacitor for the circuit.
2. What is the difference between the 3 dB frequency of a filter and its cutoff frequency?
3. Give an example of the common use of a band-pass filter.
4. If you need a factor of 80 attenuation for a frequency that is 10 times the 3-dB frequency, what order filter would you consider? (HINT: 80 = _____ dB)
5. If a plot of an amplifier circuit shows a decrease of 12 dB per factor of 2 in frequency, this circuit is acting as a filter with how many poles?
6. Design a 2-pole RC low-pass filter with a 3 dB frequency of 10 kHz using 0.001-μF capacitors. (HINT: Use scaling on the filter in Figure 1–1.)
7. Determine the magnitude response of Problem 6 by changing the frequencies of the magnitude response of the original 2-pole filter provided in Figure 1–6.
8. Chapter 12 of reference 7 contains the circuit shown in Figure 1–12, which is a normalized Chebyshev 2-pole low-pass filter with specified input and output impe-

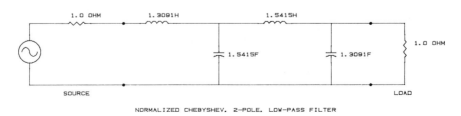

NORMALIZED CHEBYSHEV. 2-POLE. LOW-PASS FILTER

FIGURE 1–12

dances (just resistances in this case). Design a 2-pole low-pass filter with a 3-dB frequency of 15.14 kHz (95.1 krad/s). (Note that the circuit shown is normalized to have a 3-dB frequency of 1 rad/s.)

9. Use a computer program to analyze the amplitude and phase response of the original circuit of problem 8 and of the circuit you designed.

10. Redraw the circuit you designed to make it a 2-pole high-pass filter with the same 3 dB frequency. (HINT: This can be accomplished by changing the positions of the capacitors and inductors using a principle known as duality and does not require additional calculations.)

2

RADIO-FREQUENCY (RF) COILS AND TRANSFORMERS

INTRODUCTION

The use of RF coils and transformers is a requirement when working with frequencies from the commercial AM radio band (540 to 1600 MHz) through the commercial FM band (88 to 108 MHz) and up. The purpose of this experiment is to familiarize you with calculating, winding, and measuring the inductance of RF coils and of winding, measuring, and calculating the characteristics of RF transformers.

After completing this experiment you will be able to

1. Wind, calculate, and measure the characteristics of an RF coil.
2. Wind, calculate, and measure the characteristics of an RF transformer.

REFERENCES

1. Young, Chapter 1.
2. Roddy and Coolen, Chapter 1.
3. Miller, Chapter 1.
4. Tomasi (*Fundamentals*), Chapters 1 and 2.
5. Killen, Chapter 2.
6. Adamson, Chapters 1 and 2.

MATERIALS OR SPECIAL INSTRUMENTATION

#22 coated coil/transformer wire

1/4-in. dowel rod (a pencil shaft will suffice)

Toroidal coil forms

Q-meter

Quick-setting glue or cement (if a permanent transformer/coil is desired)

THEORETICAL BACKGROUND

At RF frequencies tuned circuits are required for oscillators and amplifiers. A *tuned circuit* is a combination of *LC* elements that provide a band-pass or band-stop characteristic. A band-pass characteristic enables only the frequency range of interest to be amplified or otherwise processed in the device. A wide range of capacitors and inductors are available, but the most common inductors are *chokes*. Chokes are inductors, typically with low *Q*s, used to block high-frequency signals. Most often, chokes are seen between the power supply and the collector of a transistor to ensure that RF energy is not fed through the power supply to ground, which would produce oscillation at a particular frequency. Chokes are not viable for band-pass and band-stop filters used at RF frequencies because, generally, it is desirable to work with a narrow band of frequencies for which a high-*Q* coil is required.

Coils

RF coils with specific values and adjustable coils can be purchased; however, if the desired value is not readily available, it is not difficult to wind high-*Q* coils (for the AM and FM frequency ranges with which you will work in these experiments) with values of up to about 100 μH. You will work with both solenoid and toroid coils. A *solenoid* is a coil wound around a straight cylindrical form (like a dowel rod, pencil, or ferromagnetic form). A *toroid* is a coil wound around a small donut core form. You will generally know the desired inductance (as in the RF amplifier, the FM modulator, and the FM receiver you will construct) and must wind a coil to meet the requirement.

The inductance of a coil is, theoretically, directly proportional to the square of the number of turns. The number of turns is the critical factor once you have selected your coil form. You will use two formulas for winding the desired coils in this experiment and in the later RF experiments.

Equation 2–1 is used for solenoid-type coils.

$$L \text{ (solenoid)} = N^2 \frac{R^2}{9R + 10X} \quad \text{(microhenries)} \qquad (2\text{–}1)$$

where
N = the total number of turns of wire

R = the radius of the coil in inches

X = the length of the coil in inches

Note that X is coil length, not the length of the wire making up the coil. Note also that this formula is approximate and will provide a value within about 20 percent of the desired value because of measurement errors, winding errors, and the like. This formula is used for an air, wood, or any nonferromagnetic material core.

Equation 2–2 is used for toroidal coils.

$$L \text{ (toroid)} = 0.0039N^2 \frac{D^2}{L} \quad \text{(microhenries)} \qquad (2\text{–}2)$$

where

N = the total number of turns of wire

D = the diameter of the toroid

L = the circumference of the center of the toroid

Note that this formula is for a nonferromagnetic core. Most toroid core forms have a high permeability (100 and over), and far fewer turns are required. The specification sheet, if available, will give you the number of turns required for a particular inductance. You can use this and the L-to-N^2 relationship (L is proportional to N^2) to determine the approximate number of turns required for your application.

The *quality factor* (Q) of a coil is defined by Equation 2–3.

$$Q = \frac{wL}{R} \qquad (2\text{--}3)$$

where

w = the angular frequency

L = the inductance

R = the resistance of the coil

Although it appears that Q will increase indefinitely as the frequency is increased, this is not the case because as frequency is increased, the resistance of the wire making up the coil increases due to the *skin effect*. The skin effect is due to the electromagnetic properties of current; at higher frequencies it causes more current to flow at the outer edge (the skin) of the conductor. The higher the frequency, the higher the percentage of current flow in a narrow region at the edge of the conductor (and no current flows in the majority area at the center of the conductor). This high current at the edge of the conductor (within a very small area of the conductor) increases the conductor's effective resistance as the frequency increases. Thus, Q increases with frequency until the skin effect takes effect and then decreases for very high frequencies. The coils you will use in these experiments should have a high enough Q through the commercial FM band.

RF Transformers

Two coils that have some of their magnetic force lines linked are said to be *coupled*. You are familiar with transformers used at lower frequencies, such as the standard 60-Hz transformers in most commercial power supplies. RF transformers must operate at a much higher frequency. RF coils cannot use the large amount of ferromagnetic material ("iron") necessary for complete coupling of the primary and secondary coils because this ferromagnetic material produces high losses at high frequencies.

The mutual inductance (M) between two coils is defined in Equation 2–4.

$$M = K(L_p L_s)^{1/2} \quad \text{(henries)} \qquad (2\text{--}4)$$

where

K = the coefficient of coupling

L_p = the inductance of the primary coil

L_s = the inductance of the secondary coil

K is directly related to the percentage of primary and secondary flux lines that are linked. If all flux lines are linked (as in a 60 Hz power transformer), to a good approximation, $K = 1$. If half of the flux lines are linked, $K = 1/2$, and so forth.

17

The importance of M is that it is directly related to the impedance coupled between the primary and secondary circuits. The impedance coupled into the primary from the secondary is given by Equation 2–5.

$$Z_r = \frac{(6.283FM)^2}{Z_s}$$ (2–5)

where

Z_r = the reflected impedance

F = the frequency

M = the mutual inductance

Z_s = the impedance of the secondary

Since Z_s is a complex impedance, an inductive secondary will be reflected into the primary as a capacitive impedance and vice versa.

The coupling coefficient can easily be measured for any transformer; thus, M can be calculated. To calculate the coupling coefficient, Equation 2–6 is used.

$$K = \left(1 - \frac{L_{ps}}{L_{po}}\right)^{1/2}$$ (2–6)

where K = the coefficient of coupling

L_{ps} = the inductance of the primary with the secondary short-circuited

L_{po} = the inductance of the primary with the secondary open-circuited

PROCEDURE AND QUESTIONS/PROBLEMS

Procedure

(NOTE: You need two 4-μH coils in Experiment 11 and can use the RF coil made here.)

Toroids are the primary RF cores used in industry. Note that for the 150 nH (or so) coils that you will use in the RF amplifier experiment, several turns of wire wound on a dowel rod form (or pencil) and then removed from the form will suffice.

RF Coil (Solenoid)

1. Use a 1/4-in.-diameter wooden dowel rod (or a pencil, if necessary) and #22 coated wire to wind a 4 μH inductor using the number of turns calculated from Equation 2–1.
2. Measure the inductance of the coil and comment on the difference between the calculated value and the measured value. (The lacquer coating can readily be burned off the ends using a match or lighter so that electrical contact can be made.)
3. By increasing or decreasing the number of turns, make the inductor as close to 4 μH as possible. Comment on how close you were able to get to the desired 4 μH.
4. Remove the shaft and verify that the inductance did not change (other than due to your compression or stretching of the coil).
5. Measure the Q of the inductor. If your instrumentation permits, measure the Q of the inductor at several frequencies and comment on the differences, if any.
6. Keep the 4 μH coil constructed for use in Experiment 11.

RF Transformer (Toroid)

1. Using a toroidal coil form with a high permeability (ferromagnetic material), wind a 50 μH coil on one side (this will be called the primary) and a 10-μH coil on the other side (secondary) of the "donut."
2. Adjust these to be as close to 50 μH and 10 μH as possible.
3. Short-circuit the secondary leads (the 10-μH). Remeasure the inductance of the primary coil. Calculate the coupling coefficient K from Equation 2–6.
4. Comment on the value of K versus the coupling coefficient for an ideal transformer.

Questions/Problems

1. Calculate M, the mutual inductance of your transformer.
2. If an impedance of $50 - j50$ (series RC) is attached to the secondary, find the impedance reflected back to the primary (use Equation 2–5).
3. Explain how your transformer could be used to match the impedances of two devices.

3

OSCILLATORS

INTRODUCTION

You have met and experimented with oscillator circuits in previous electronics courses. The oscillator circuits you will build in this experiment will demonstrate additional principles used extensively in communications. Oscillations can be a hindrance if you are trying to build an amplifier or a help if you need a specified frequency for modulation. Many integrated circuit oscillators and function generators are available that provide a desired output frequency (sinusoidal or square wave) with the addition of a few components (the 555 timer chip is a well-known example). The objective of this experiment is to provide insight into how oscillators operate.

First, you will construct a phase-shift oscillator. This will allow you to observe the effect of positive feedback on an amplifier. Then you will build a Pierce crystal oscillator and observe the stability of this type of oscillator. Crystal oscillators are required for almost all communications circuits.

After completing this experiment, you will be able to

1. Understand the reasons for amplifier instability or oscillation due to feedback.
2. Construct and use a Pierce crystal oscillator.

REFERENCES

1. Young, Chapter 2.
2. Roddy and Coolen, Chapter 6.
3. Miller, Chapter 1.

4. Tomasi (*Fundamentals*), Chapter 2.
5. Killen, Chapter 7.
6. Adamson, Chapter 15.

MATERIALS OR SPECIAL INSTRUMENTATION

RC Phase-Shift Oscillator

Devices: 741 op amp

Resistors: 100 k potentiometer, three 1k

Capacitors: three 0.1 μF

Pierce Crystal Oscillator

Devices: 2N5457 (or MPF 102) *n*-channel JFET with high g_m
3.579 MHz colorburst crystal

Resistors: 10 k potentiometer

Capacitors: 100 pF, 50 pF, two 0.1 μF

Inductors: 100 μH

THEORETICAL BACKGROUND

The basic principle used in all oscillators (intentionally built or produced due to inadvertent feedback in an amplifier) is positive feedback from the output back to the input at a particular frequency. When you learned to build a single-stage, common-emitter transistor amplifier, you used a bypassed emitter resistor, such as R_e shown in Figure 3–1, to stabilize the dc operating point. This emitter resistor provided *negative* feedback of the dc voltage across R_e back to the input. Negative feedback implies a 180° phase shift from the output back to the input.

   Positive feedback is undesirable in an amplifier because it causes oscillations, and then you no longer have an amplifier (at a particular frequency). If you need an oscillator, this positive feedback is required. To be more specific, the following conditions (known as the *Barkhausen criteria*) must be fulfilled to produce an oscillator:

1. The feedback signal must have a 360° (or 0°) phase shift measured from the input through the output and back along the feedback path to the input.

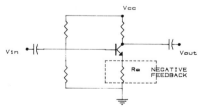

TRANSISTOR AMPLIFIER WITH NEGATIVE FEEDBACK

FIGURE 3–1

22

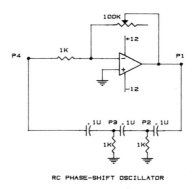

RC PHASE-SHIFT OSCILLATOR

FIGURE 3–2

2. The total gain measured from the input through the output and back along the feedback path to the input must be 1.

RC Phase-Shift Oscillator

The *RC* phase-shift oscillator shown in Figure 3–2 illustrates the preceding criteria. With the lower feedback loop removed, the circuit is a simple amplifier. Figure 3–3 shows the gain and phase response of the 3-pole filter of the *RC* phase-shift oscillator.

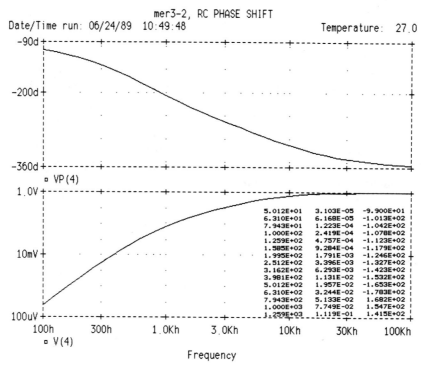

FIGURE 3–3

The formula for the 3 dB frequency of the 3-pole RC high-pass filter used is given in Equation 3–1.

$$f = \frac{1}{15.39RC} = 650 \text{ Hz} \tag{3–1}$$

Using Figure 3–3 (and the associated printout), you can see that the three RC sections of the high-pass filter provide an overall phase shift of 180° at approximately 650 Hz. Note that individual sections of the filter do *not* contribute 60° each due to interaction between sections. You can also see that the gain of the filter at 650 Hz is about -30 dB. The Barkhausen criteria require a 360° phase shift and a gain of $+1$. The inverting op-amp configuration provides a 180° phase shift, and at 650 Hz, the RC network provides the other 180°. The loss of 30 dB must be made up by the gain of the op amp. Solving 20 $\log_{10} X = 30$ provides a required gain of 31.6 for the amplifier to make the loop gain through the RC network $+1$. Building a feedback oscillator is as simple as ensuring 360° and a gain of 1 in the feedback path at the desired oscillation frequency.

Pierce Crystal Oscillator

Figure 3–4 is a Pierce crystal oscillator using a 3.5789 MHz colorburst crystal. The 0.1 μF capacitor in the feedback path blocks the dc but is a short circuit to ac, so an ac feedback path is available from the output to the input of the amplifier. The 100 μH inductor blocks the oscillator frequency from the power supply. The 50 pF capacitor provides a short circuit to ground for harmonics of the basic 3.579 MHz frequency and provides an output that is more sinusoidal. The 100 k resistor provides a dc path to ground for self-biasing of the FET. The 100 pF capacitor provides a load of $-j2800$ (at 3.579 MHz). So the impedance is relatively high (for high gain), and an additional phase shift is introduced at the drain of the FET.

Note that to meet the Barkhausen criteria, the voltage gain of the JFET at 3.579 MHz must be relatively large, so the JFET used must have a relatively high g_m. If the ac-coupled load is too large (too small a resistor equals too large a load equals high current output), the oscillator cannot sustain oscillations because the gain is reduced. A crystal oscillator using a discrete device (here a JFET) is not difficult to construct. Many crystal-oscillator-based circuits use integrated circuits, in which the manufacturer has specified the crystal location and accompanying capacitors. This makes oscillator construction even more straightforward.

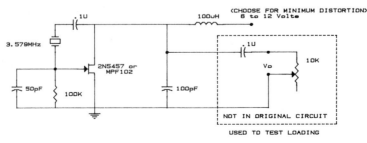

PIERCE CRYSTAL OSCILLATOR

FIGURE 3–4

24

PROCEDURE AND QUESTIONS/PROBLEMS

Procedure

RC *Phase-Shift Oscillator*

1. For Figure 3–2 measure and record all resistor and capacitor values.
2. Construct the circuit with the potentiometer at a low value (0 to 10 kΩ).
3. Increase the value of the op-amp gain until oscillations first occur. Record the resistance of the pot and the frequency of the oscillations at this point. Comment on the difference between the measured and calculated gain for oscillation and the frequency of oscillation (Equation 3–1).
4. Continue increasing the resistance of the pot and note that oscillations continue but that the frequency of the oscillations decreases. Explain this phenomenon.
5. *Readjust the pot to the point that oscillations just begin.* Measure and record the phase shift (use a 10:1 probe for minimum interference from the oscilloscope) at points P_1, P_2, P_3, and P_4. Explain what is happening as the sinusoidal oscillation is filtered through the RC network.
6. Measure the amplitude and frequency of the oscillations at point P_1 and the amplitude of the oscillations at point P_4 and record. Explain the difference in amplitude.

Pierce Crystal Oscillator

1. Measure all component values and record.
2. Construct the circuit [*note that some MPF 102s do not have sufficient gain to sustain oscillations, so you may have to try several MPF 102s or use an n-channel JFET with higher gain* (such as the 2N5457)] and measure the output voltage, frequency, and phase shift at V_o. Measure the amplitude and phase shift at the input to the JFET. Record these measurements. Explain how feedback occurs in the oscillator.
3. Remove the 50 pF bypass capacitor. Discuss the change in the output waveform. Reinsert the 50 pF bypass capacitor.
4. Place a 50 pF capacitor in parallel with the crystal. Determine and record the new oscillation frequency. Explain this frequency change.
5. Remove the 50 pF capacitor in parallel with the crystal and place it in series with the crystal. Determine and record the new oscillation frequency. Remove the 50 pF series capacitor. Explain this frequency change.
6. Using the original circuit, add the 0.1 μF capacitor and the 10 kΩ potentiometer (initially set at its maximum resistance) as a load. Decrease the resistance of the pot until oscillations cease. Record this value of resistance. Discuss the types of active devices that could be attached to the oscillator based on this measurement. (HINT: Consider high/low impedances.)

Questions/Problems

1. Design a phase-shift oscillator to provide a frequency of 1 kHz. Analyze your design using a computer analysis program. Discuss the effect of an oscilloscope probe on the circuit. Discuss the Barkhausen criteria as they apply to this circuit.
2. Design a Pierce crystal oscillator that uses a 5.1 MHz crystal. Discuss how the Barkhausen criteria apply to your design.

3. Design a circuit that would provide a buffer between the output of the 3.579 MHz oscillator in Figure 3–4 and a mixer diode.
4. Discuss how you could obtain 7.1578 MHz (twice the crystal frequency) from the circuit of Figure 3–4. Should the 50 pF bypass capacitor be used if you are trying to extract the 7.158 MHz harmonic? (HINT: Harmonics are produced by nonlinearities.)

4

FOURIER SERIES AND FILTERS

INTRODUCTION

The Fourier series representation of periodic time-domain functions is one of the fundamental methods of relating the time and frequency domains. In Experiment 1 you studied the characteristics of filters in the frequency domain. In this experiment you will observe the effect of a 2-pole low-pass filter on a squarewave input and see how Fourier series can be used to predict the output using ac (complex) analysis methods that you used in beginning circuit theory.

   After completing this experiment, you will be able to

1. Determine the Fourier series components of a squarewave or a ramp.
2. Determine the transfer function of a passive filter.
3. Find the Fourier components of the output waveform given the Fourier components of a periodic waveform and the transfer function of a device.
4. Find the time-domain representation from the Fourier components of an output waveform.

REFERENCES

1. Young, Chapter 3.
2. Roddy and Coolen, Chapter 2.
3. Miller, Chapter 1.
4. Tomasi (*Fundamentals*), Chapter 1.
5. Killen, Chapter 4.
6. Adamson, Chapter 3.
7. TKSOLVERPLUS, Universal Technical Systems, Inc., Rockford, Ill., 1988.

MATERIALS OR SPECIAL INSTRUMENTATION

Spectrum analyzer (if available)

Equation solver

Resistors: two 5 k potentiometers

Capacitors: two 0.001 μF

THEORETICAL BACKGROUND

You learned the basic principles of frequency-domain analysis in your first year ac circuits course when you used phasor notation to determine the output of a circuit to a given sinusoidal input. An example of such analysis is finding the response of a single-pole low-pass filter to an input sinusoid.

An example using the circuit of Figure 4–1 is provided here for review: Since a capacitor is represented by $1/jwC$, simple voltage division can be used to develop the transfer function of the filter, which is shown in Equation 4–1.

$$\frac{V_{out}}{V_{in}} = \frac{1/jwC}{R + 1/jwC} = \frac{1}{1 + jwRC} \qquad (4\text{--}1)$$

Given values for R and C and a sinusoidal (phasor) input, the sinusoidal (phasor) output can be found as an amplitude and phase, as shown in Equations 4–2 and 4–3.

$$|V_{out}| = |V_{in}| \times [1 + (wRC)^2]^{-1/2} \qquad (4\text{--}2)$$

where the notation $|V|$ means the absolute value, or amplitude.

$$\angle V_{out} = \angle V_{in} - \arctan(wRC) \qquad (4\text{--}3)$$

where the notation $\angle V$ means angle V.

For instance, for $R = 1 \text{ k}\Omega$ and $C = 1 \mu$F, with an input of $5 \sin(6.283 \times 200t + 45°)$, the following calculations apply. First, find the phasor representation of the sinusoidal input. Note that the peak value of the sinusoid is used here and that the *sine* is used as the base waveform. In some texts the rms value of the sinusoid is used (this is necessary if calculating power), and in other texts the cosine is used as the base. Either choice is arbitrary, since the phasor notation is only a representation of the sinusoidal waveform that simplifies the calculations. Thus, the input waveform is represented by

$$V_{in} = 5\angle 45°, \qquad w = 6.283 \times 200 = 1.2566 \times 10^3 = 1256.6$$

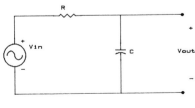

SINGLE POLE. RC. LOW–PASS FILTER

FIGURE 4–1

The angular frequency, R and C, is then substituted into Equations 4–2 and 4–3 and the amplitude and phase are calculated.

$$|V_{out}| = 3.114$$

$$\angle V_{out} = \angle -6.5°$$

This must now be converted back into the time domain to find the output that would be observed on an oscilloscope.

$$V_{out} = 3.114 \sin(6.283 \times 200t - 6.5°)$$

Notice (and recall) that only the amplitude and the phase have changed, while the frequency remained the same.

Fourier Series and Filters

The Fourier series representation of a square wave (assuming odd symmetry, amplitude $\pm V$, and 0 dc level) is shown in Equation 4–4. The Fourier series for a triangular wave (assuming even symmetry, amplitude $\pm V$, and 0 dc level), is shown in Equation 4–5. The theory of Fourier series is covered in most communications textbooks and will not be repeated here. Notice, for the purpose of this experiment that a Fourier series is a summation of an infinite series of single-frequency sinusoids. Also note that the amplitude of higher frequency terms decreases much more rapidly for the triangular waveform than for the squarewave, with its very sharp edges.

$$v(t) = 1.2732V \left[\sin(wt) + \left(\frac{1}{3}\right) \sin(3wt) + \left(\frac{1}{5}\right) \sin(5wt) + \left(\frac{1}{7}\right) \sin(7wt) \right.$$

$$\left. + \left(\frac{1}{9}\right) \sin(9wt) + \left(\frac{1}{11}\right) \sin(11wt) + \cdots \right] \text{(square)} \tag{4-4}$$

$$v(t) = 0.4053V \left[\cos(wt) + \left(\frac{1}{9}\right) \cos(3wt) + \left(\frac{1}{25}\right) \cos(5wt) \right.$$

$$\left. + \left(\frac{1}{49}\right) \cos(7wt) + \left(\frac{1}{81}\right) \cos(9wt) + \cdots \right] \text{(triangular)} \tag{4-5}$$

The two-pole low-pass filter to be used in the experiment is shown in Figure 4–2. The amplitude and phase response of the filter, the program used, and a printout of values are shown in Figure 4–3. This filter was scaled (for a 50 kHz, 3 dB point and to use 0.001 μF capacitors) from the two-pole low-pass filter used in Experiment 1.

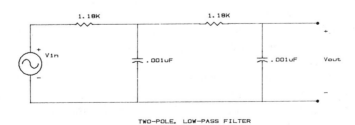

TWO-POLE. LOW-PASS FILTER

FIGURE 4–2

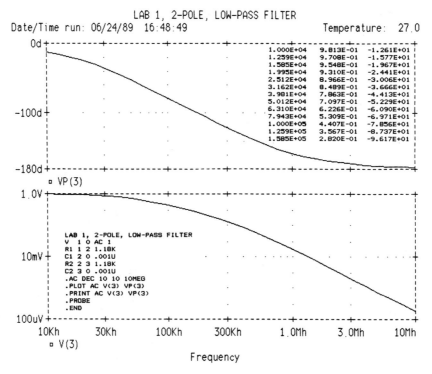

FIGURE 4–3

Analysis in the Frequency Domain

Much of the analysis performed in communications (and in control systems) is done using the frequency domain. Use of Fourier series to determine the output time-domain waveform when a square or ramp function is used as the input to a filter provides an introduction to these methods of analysis.

The theory behind Fourier series analysis of an output given any periodic waveform input to a circuit is conceptually easy. Each frequency component of the input is converted to phasor notation, the complex transfer function of the circuit is found, and each individual frequency component of the Fourier series input is multiplied by the transfer function. Each individual frequency component of the output is converted back into a sinusoidal representation (from phasor notation), and the infinite sum of the output sinusoids is plotted to determine the output. Of course, this is very tedious if done by hand, but computers can readily perform these tasks. Additionally, you cannot add an infinite sum of terms, as in all Fourier series, but you can add sufficient terms to closely approximate the input and the output waveforms.

For an initial calculation, look at a 10-kHz squarewave input to the filter in Figure 4–2 and determine the output waveform by Fourier series analysis. Assume the input to be from 0 to 4 V in amplitude and to be an odd waveform. Find the Fourier series for this input by substituting $V = 4$ and $w = 6.283 \times 10^4$ in Equation 4–4 and adding 2 V dc to make the waveform from 0 to 4 V instead of -2 to $+2$ V. The resulting Fourier series is shown in Equation 4–6.

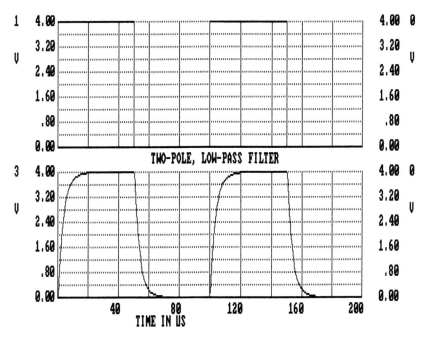

TWO-POLE, LOW-PASS FILTER

TIME IN US

FIGURE 4–4

$$v(t) = 2 + 2.546 \sin(wt) + 0.849 \sin(3wt) + 0.509 \sin(5wt)$$
$$+ 0.364 \sin(7wt) + 0.283 \sin(9wt) + 0.231 \sin(11wt) \qquad \textbf{(4–6)}$$
$$+ 0.196 \sin(13wt) + \cdots$$

After a bit of calculation, the transfer function of the two-pole low-pass filter shown in Figure 4–2 is given by Equation 4–7.

$$\frac{V_{\text{out}}}{V_{\text{in}}} = \frac{7.184 \times 10^{11}}{(7.184 \times 10^{11} - w^2) + jw(2.542 \times 10^6)} \qquad \textbf{(4–7)}$$

This can be converted to a phasor at each individual frequency, which entails many calculations because each phasor will have an amplitude and phase component. The response of this filter to the described squarewave input is shown in Figure 4–4. Calcu-

TABLE 4–1

Frequency (kHz)	Amplitude	Phase (degrees)
10	0.981	−12.6
30	0.861	−35.1
50	0.711	−52.2
70	0.582	−64.9
90	0.482	−74.5
110	0.405	−82.2
130	0.345	−88.6

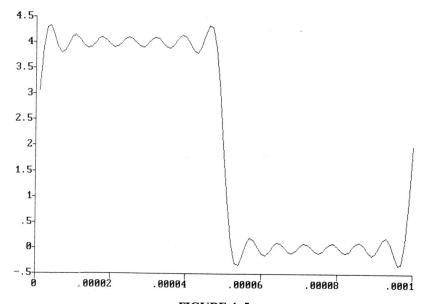

FIGURE 4-5

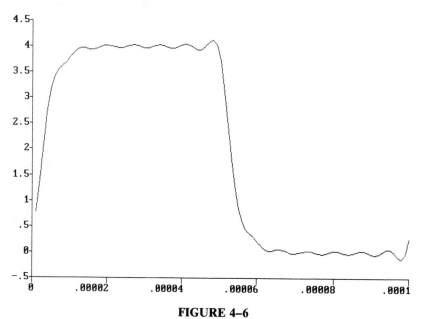

FIGURE 4-6

lating the phasor response of Equation 4–7 at the appropriate frequencies provides you with amplitude and phase information (up to 130 kHz) at each frequency shown in Table 4–1. Multiplying each amplitude and adding each phase to the original input provides the Fourier series representation of the output shown in Equation 4–8.

$$v(t) = 2 + 2.5 \sin(wt - 12.6°) + 0.731 \sin(3wt - 35.1°)$$
$$+ 0.362 \sin(5wt - 52.2°) + 0.211 \sin(7wt - 64.9°)$$
$$+ 0.136 \sin(9wt - 74.5°) + 0.0938 \sin(11wt - 82.2°) \qquad \textbf{(4–8)}$$
$$+ 0.0678 \sin(13wt - 88.6°)$$

An equation plotter (see reference 7 at the beginning of the chapter), is used to create a graph of the band-limited (the Fourier series is plotted through only 130 kHz) input squarewave (from Equation 4–6), as shown in Figure 4–5. A graph of the output (Equation 4–8) is shown in Figure 4–6 (reference 7). Comparison of these Fourier representations with Figure 4–4 reveals that the Fourier series representation approaches the actual output (and would be the same except that you did not plot an infinite number of frequencies) calculated using a network analyzer.

Fourier series (and related Fourier transform) methods are widely used in analysis (using computers, of course), and the method just shown provides a good introduction to calculation of output using the frequency domain.

PROCEDURE AND QUESTIONS/PROBLEMS

Procedure

1. Measure the resistance and capacitance of all components.
2. Construct the 2-pole low-pass filter shown in Figure 4–2.
3. Verify that the output amplitude and phase shift of the filter match the calculated values with a 10-, 30- and 50-kHz, 1-V peak, sine-wave input.
4. Verify that the output with a 10-kHz, 0- to 4-V squarewave input matches that of Figure 4–4.
5. Observe the output with a 0- to 4-V, 10-kHz ramp input.

Questions/Problems

1. Determine the first four terms in the Fourier series representation of a 0- to 4-V, 10-kHz ramp waveform.
2. Calculate the phasor transfer function of the filter used in lab for a 75-kHz input sine wave.
3. Determine the first four terms in the Fourier series representation of the output of the filter for a 0- to 4-V, 10-kHz ramp input. (HINT: Use Equation 4–7.)
4. Why are fewer Fourier series terms required to represent the ramp waveform than the squarewave?
5. Use a computer program to plot the first four terms of the input and output for a ramp waveform, using the filter shown.

5

SINGLE-TRANSISTOR MIXER WITH BAND-STOP AND LOW-PASS FILTER

INTRODUCTION

The fundamental principle of modulation involves the mixing or multiplying of a low-frequency signal with a higher frequency signal, such as an AM or FM carrier. This enables the information contained in the low-frequency signal to be transmitted through space as high-frequency electromagnetic waves. Commercial radios use an intermediate frequency (IF) (455 kHz for commercial AM, 10.7 MHz for commercial FM, and 70 MHz for satellite receivers are commonly used). These IF carriers contain all the information available to the receiver, but in order to obtain this information, mixing must take place to obtain the lower-frequency signals ''riding'' on the IF carrier. This principle is also used in mixing low-frequency signals up to IF or RF frequencies. The purpose of this experiment is to observe the effect of mixing two frequencies using a nonlinear, single-transistor mixer and to demonstrate the use of a band-stop and a 2-pole low-pass filter in a practical application.

After completing this experiment, you will be able to

1. Predict the frequencies generated in a nonlinear mixer.
2. Design and build a twin-T band-stop filter.
3. Understand and analyze the combined effect of two filters.
4. Use a single-transistor mixer for modulation or demodulation.

REFERENCES

1. Young, Chapter 7.
2. Roddy and Coolen, Chapters 7 and 8.

3. Miller, Chapters 2 and 3.
4. Tomasi (*Fundamentals*), Chapters 2 and 3.
5. Killen, Chapter 8.
6. Adamson, Chapters 4 and 5.
7. *Reference Handbook for Radio Engineers.*

MATERIALS OR SPECIAL INSTRUMENTATION

Spectrum analyzer (if available)

Two signal generators

Devices: 2N3904 *npn* transistor

Resistors: two 27 k, 22 k, two 10 k, two 10-k potentiometers, 4.7 k, 3.2 k, 100 Ω

Capacitors: two 1 μF, 0.1 μF, three 100 pF, two 50 pF

THEORETICAL BACKGROUND

Mixing is the nonlinear combination of two signals to produce sum and difference frequencies and harmonics of the signals. It is primarily used for modulation or demodulation of a signal. *Modulation* (in this usage of the term) is the translation of the signal information to a higher frequency signal, and *demodulation* is the translation of signal information carried by a high-frequency signal down to a lower frequency. The principle involved in this experiment is the use of the nonlinear portion of a transistor characteristic to mix (combine) two signals.

All the mathematical functions we normally work with can be represented by a *Taylor series,* which is a summation of increasing powers of the independent variable. Equation 5–1 is an example of a Taylor series for the exponential function (which can match diode and transistor characteristics over some range).

$$e^t = 1 + t + \frac{t^2}{2!} + \frac{t^3}{3!} + \frac{t^4}{4!} + \cdots \tag{5–1}$$

You have already seen that any periodic function can be represented by a Fourier series that consists of sinusoids and can be thought of as a discrete frequency spectrum (discrete meaning individual terms at a fundamental frequency and harmonics). Nonperiodic functions can also be represented by a frequency spectrum using Fourier transform techniques (not covered here but available in many computer programs). Since frequencies are just sinusoidal terms, if a single-frequency term is introduced into a nonlinear circuit that is approximately exponential, the result is the square, the cube, and so forth of the sinusoid, as shown by Equation 5–1. From trigonometry, Equation 5–2 shows the result of squaring and cubing a single sinusoid (the cosine function is used here, but the same result is obtained using a sine function) as the result of nonlinear combination.

$$e^{\cos(\omega t)} = 1 + \cos(\omega t) + 0.5\cos^2(\omega t) + 0.167\cos^3(\omega t) + \cdots$$
$$= 1 + \cos(\omega t) + 0.25[1 + \cos(2\omega t)] + \text{(higher-order terms)} \tag{5–2}$$

Notice that a squared term produces a harmonic at twice the original frequency and a cubed term produces a harmonic at three times the original frequency. This characteristic of nonlinear devices of producing a signal rich in harmonics can be both good and bad in a circuit. Harmonics are undesirable if you want a linear amplifier such as in audio amplification and don't want any harmonics polluting the signal. If you want to generate higher frequencies or mix frequencies, as you will do in this experiment, harmonics are necessary.

If two signals are mixed in a nonlinear device, the result is a signal that is very rich in harmonics. If you provide two input sinusoids to a nonlinear device such as the exponential represented in Equation 5–1, the output contains all possible combinations of the original signals. For instance, Equation 5–3 (which is a substitution of the exponent into Equation 5–1) illustrates the output when only the terms through the square term are multiplied. The two angular frequencies considered as inputs are ω and m.

$$
\begin{aligned}
e^{\cos(\omega t)\cos(mt)} &= 1 + \cos(\omega t)\cos(mt) + [\cos(\omega t)\cos(mt)]^2 \\
&\quad + [\cos(\omega t)\cos(mt)]^3 + \cdots \\
&= 1 + 0.5[\cos(\omega + m)t + \cos(\omega - m)t] \\
&\quad + \{0.25 + 0.125\cos[2(\omega + m)t]\} + 0.125\cos[2(\omega - m)t] \\
&\quad + 0.25\cos(2\omega t) + 0.25\cos(2mt) \\
&\quad + \text{(higher-order terms from the cube, } \ldots)
\end{aligned}
\tag{5-3}
$$

Figure 5–1 shows the circuit you will use in this experiment. The V_y input is the local oscillator and is made large (several volts), so that the signal is cut off at the collector. This forces the transistor to operate in a very nonlinear region. (This region is not necessarily exponential, but nonlinearities generate harmonics similar to the exponential example.) Now that the transistor is operating in a nonlinear region during part of its cycle, the input signal, representing an input from the IF of a radio, is introduced at a low level (millivolts). The result is an output at the collector of the transistor that contains all the frequencies in Equation 5–3 and many higher harmonics and sum and difference frequency harmonics due to the nonlinearity.

As an example, if V_y is 500 kHz and V_x is 520 kHz, the output frequencies at the collector (from Equation 5–3) will be dc, 20 kHz, 40 kHz, 1.00 MHz, 1.02 MHz, 1.04 MHz, 2.04 MHz, and other harmonics of the sum frequencies, the difference frequencies, and the original frequencies. In a radio you are interested only in the audio output; therefore, you need to eliminate all higher frequencies. In this circuit you are interested

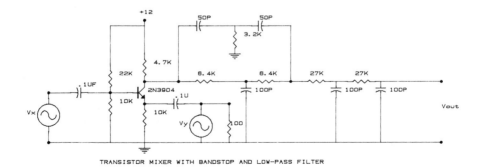

TRANSISTOR MIXER WITH BANDSTOP AND LOW-PASS FILTER

FIGURE 5–1

37

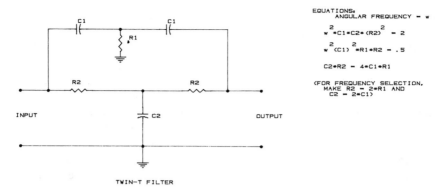

EQUATIONS:
ANGULAR FREQUENCY = w

$$w^2 *C1*C2*(R2)^2 = 2$$

$$w^2 (C1)^2 *R1*R2 = .5$$

$$C2*R2 = 4*C1*R1$$

(FOR FREQUENCY SELECTION,
MAKE R2 = 2*R1 AND
C2 = 2*C1)

TWIN-T FILTER

FIGURE 5–2

only in the 20-kHz signal (which is above the usual audio range but provides a good signal on many spectrum analyzers), so you must design a filter to eliminate all the higher frequency signals. The 500 kHz component is the largest in the output because of the very large signal generated by the local oscillator (V_y). Thus, a special notch filter will be used to attenuate this large signal, and a low-pass filter will be used to attenuate other harmonics.

Figure 5–2 shows the twin-T notch filter for the circuit and the applicable equations. Note that the notch rapidly becomes more shallow, and therefore less effective, as you deviate from the calculated values.

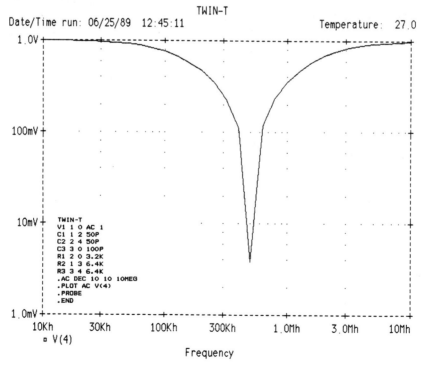

FIGURE 5–3

Figure 5–3 shows the frequency (amplitude) response of the twin-T filter used in the experiment. Note that the notch was calculated at 500 kHz and provides good attenuation for the large 500 kHz signal component if the *R* and *C* values of the filter are as shown.

Figure 5–4 shows the frequency response of the 2-pole low-pass filter (at the output of Figure 5–1) designed for the circuit. Note that the 2-pole filter is designed to have a 3 dB frequency of 20 kHz. The filter attenuates the 20 kHz difference frequency by 3 dB (factor of 2), and the attenuation increases so that higher frequency components are more heavily attenuated. Note that a 3- or 4-pole filter would be more effective in eliminating undesirable higher frequencies.

Figure 5–5 shows the combined frequency response of the two filters as used in the circuit. Notice that because the frequency responses of the two circuits are at very different frequency ranges, the combined response is approximately the linear sum of the two separate responses. This is not true if two filters in the same frequency range are combined. The combination of all the elements in the circuit we have discussed is the mixing of two signals and the filtering (elimination) of all but the signal of interest. This principle is widely used, and the theory is applicable in many circuits of interest in communications.

This principle of mixing two frequencies to obtain sum and difference frequencies will be used in the next experiment to produce an AM signal using a special-purpose integrated circuit. The transistor mixer discussed here could also be used to mix a lower

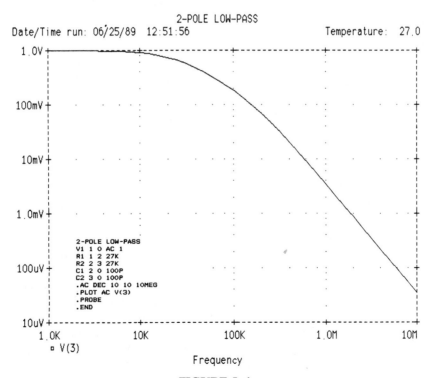

FIGURE 5–4

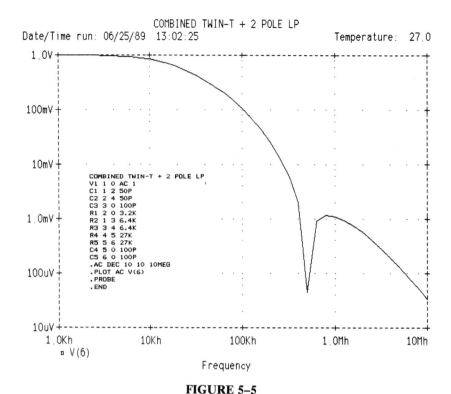

```
COMBINED TWIN-T + 2 POLE LP
V1  1  0  AC  1
C1  1  2  50P
C2  2  4  50P
C3  3  0  100P
R1  2  0  3.2K
R2  1  3  6.4K
R3  3  4  6.4K
R4  4  5  27K
R5  5  6  27K
C4  5  0  100P
C5  6  0  100P
.AC DEC 10 10 10MEG
.PLOT AC V(6)
.PROBE
.END
```

Frequency

FIGURE 5–5

frequency signal with a higher frequency signal to modulate a carrier with an audio signal by introducing the audio signal at the V_x input with the carrier at V_y.

PROCEDURE AND QUESTIONS/PROBLEMS

Procedure

1. Construct the circuit of Figure 5–1.
2. With signal V_x disconnected, set V_y at 500 kHz (this represents the local oscillator signal). Increase the amplitude of V_y until the signal at the collector is noticeably nonlinear (note that the signal is clipped). A large signal (approximately one-half of V_{CC}) will be necessary. The clipping is necessary to introduce nonlinearities, which is how the sum and difference frequencies and other harmonics are produced. Observe the harmonics on the spectrum analyzer, if one is available.
3. Set V_x at 520 kHz with a 20 to 100 mV signal. (This represents the IF signal coming in from a radio with a 20 kHz signal as its upper sideband (AM IF is 455 kHz.)
4. Observe the signal at the collector using an oscilloscope. This time-domain waveform will probably have little meaning because you are mixing frequencies and are interest in only one frequency, 20 kHz, while the many sum and difference and harmonic frequencies produce a complex time-domain waveform.
5. Observe and comment on the frequency spectrum at the collector using a spectrum

40

analyzer. List the frequencies you expected and observed at the collector, and sketch the observed frequency spectrum.

6. Observe the signal at the output of the circuit using an oscilloscope. Comment on the observed signal. Note that there is a dc bias on the output from the collector. To eliminate the bias, if necessary, use the other 1-μF capacitor provided.

7. Use a spectrum analyzer to observe the frequency spectrum at the circuit output. Sketch the spectrum.

8. Comment on the effectiveness of the filters.

Questions/Problems

1. Design a single-transistor mixer similar to the one used in this experiment to modulate a 5 kHz input onto a 1 MHz carrier.
   a. Explain how you could verify your modulated signal using an AM receiver.
   b. Sketch the frequency spectrum that you would expect at the collector of your design.
   c. Since the desired output is a 5 kHz signal on a 1 MHz carrier, how could you eliminate the 5 kHz (and other low frequencies) from the output of your design?

2. Explain how the nonlinear portion of a diode characteristic (assume that the portion of the diode characteristic used is approximately parabolic) could be used to produce harmonics of an input signal.
   a. Sketch the frequency spectrum you would expect if a 1 MHz, 5 V sine wave was input to an ideal diode.
   b. What difference would it make if the diode were considered to be exponential instead of parabolic?

3. Explain how you could produce a 3 MHz signal using a nonlinear device with only a 1 MHz signal available. Explain how you could eliminate the other harmonics and pass only the desired 3 MHz signal using filters.

4. Analyze the circuit of Figure 5–1 using a computer program.

6

AMPLITUDE MODULATION

INTRODUCTION

Amplitude modulation was the earliest form of widespread RF modulation and is the easiest to understand. In this experiment you will look at amplitude modulation of two sinusoidal waveforms. You will begin by multiplying a sinusoid by itself (squaring the sine wave), thereby doubling the frequency. Then you will use two different frequency sinusoids and produce double sideband-suppressed carrier (DSB-SC) and amplitude modulation (AM). The MC1496 balanced modulator, a monolithic IC will be used to produce the desired waveforms. The MC1496/1596 ICs are versatile and can be used up to 200 MHz.

After completing this experiment, you will be able to

1. Understand frequency doubling.
2. Understand AM and DSB-SC modulation.
3. Construct a frequency doubler, a DSB modulator, and an AM modulator using a balanced modulator IC.

REFERENCES

1. Young, Chapters 5 and 8.
2. Roddy and Coolen, Chapters 8 and 9.
3. Miller, Chapters 2 and 3.
4. Tomasi (*Fundamentals*), Chapters 2, 3, and 4.
5. Killen, Chapter 8.
6. Adamson, Chapters 4, 5, and 15.
7. TKSOLVERPLUS, Universal Technical Systems, Inc., Rockford, Ill., 1988.

MATERIALS OR SPECIAL INSTRUMENTATION

Two signal generators

Spectrum analyzer (if available)

AM radio

Devices: MC1496/1596 balanced modulator IC. The 14-pin DIP circuit is shown in Figure 6–4. If you are using the 10-pin metal can, see the Appendix for the correct pin numbers.

Resistors: three 1 k, two 3.9 k, two 10 k, three 51 Ω, 6.8 k, 50 k potentiometer

Capacitors: four 0.1 μF

THEORETICAL BACKGROUND

A balanced modulator is essentially a multiplier. The output of the MC1496 balanced modulator is proportional to the product of the two input signals. If you apply the same sinusoidal signal to both inputs of a balanced modulator, the output will be the square of the input signal. Using a common trigonometric identity, the output can be shown to contain a frequency that is twice the input frequency. This is shown in Equation 6–1 for a 10-kHz sinusoid. Note that the 10-kHz sinusoid is converted into a 20-kHz sinusoid using this multiplication process.

$$\text{Trig identity:} \quad [\cos(wt)]^2 = 0.5[1 + \cos(2wt)]$$
$$A\cos(2 \times 10,000t)B\cos(2 \times 10,000t) = AB[\cos(2 \times 10,000t)]^2 \qquad \textbf{(6–1)}$$
$$= AB[0.5(1 + \cos(2 \times 2 \times 10,000t)]$$

If you use two sinusoidal signals with different frequencies at the two inputs of a balanced modulator (multiplier) you can produce AM-DSB/SC modulation. This is generally accomplished using a high-frequency "carrier" sinusoid and a lower frequency "modulation" waveform (such as an audio signal from a microphone). Again, you may use a trigonometric identity to see, mathematically, what the output will look like. Using a 100 kHz carrier and a 5 kHz sinusoid for modulation, the mathematical analysis of the multiplication process (the output) is shown in Equation 6–2. This equation indicates that the resulting waveform contains neither of the original frequencies but only the sum and difference frequencies of the two inputs.

$$\text{Trig identity:} \quad \cos(a)\cos(b) = 0.5[\cos(a + b) + \cos(a - b)]$$
$$\cos(2 \times 100,000t)\cos(2 \times 5000t) = 0.5[\cos(2 \times 105,000t) \qquad \textbf{(6–2)}$$
$$+ \cos(2 \times 95,000t)]$$

Figure 6–1 is a plot of a DSB-SC waveform (see reference 7 at the beginning of the chapter). This figure is the graph of a 100 kHz and a 5 kHz sinusoid multiplied together and shows the result of Equation 6–2. Note from the equation that neither of the original frequencies remains in the output. The envelope formed by connecting the peaks of the sinusoids forms a distinctive waveform that looks much like a half-wave-rectified sine wave and its mirror image.

Any periodic waveform can be decomposed into an infinite series of sinusoids using Fourier series techniques. As an example, a 5 kHz triangular wave can be used as the modulation waveform. The first four terms of a Fourier series representation of a 5

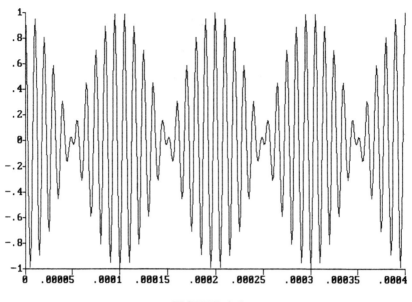

FIGURE 6-1

kHz, 1 V peak triangular wave are shown in Equation 6–3. Each of these four Fourier series terms is multiplied by the 100 kHz carrier. The result is a series of sum and difference frequencies. The overall effect of Equation 6–3 is shown in Figure 6–2.

$$v(t) = [0.405 \cos(2 \times 5000t) + 0.045 \cos(2 \times 3 \times 5000t)$$
$$+ 0.0162 \cos(2 \times 5 \times 5000t)$$
$$+ 0.00827 \cos(2 \times 7 \times 5000t)]$$
$$\times [\cos(2 \times 100,000t)]$$

(6–3)

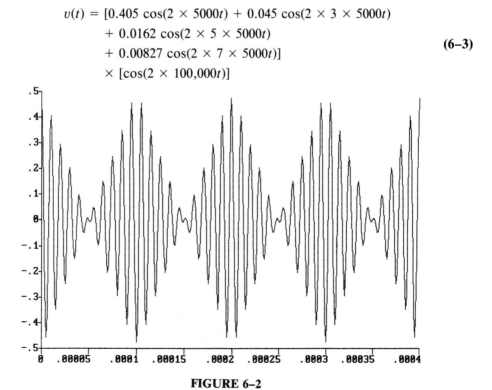

FIGURE 6-2

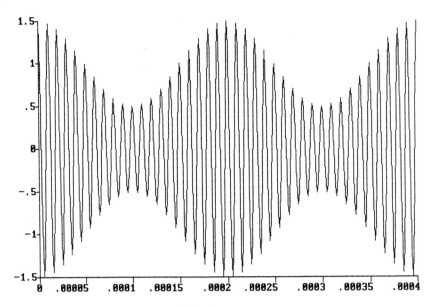

FIGURE 6-3

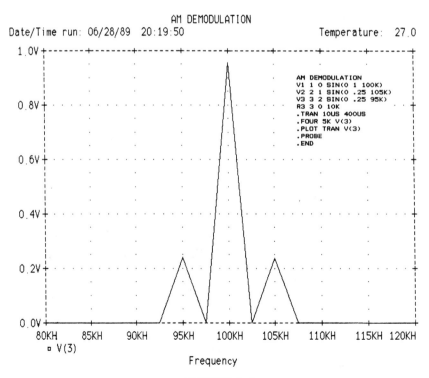

FIGURE 6-4

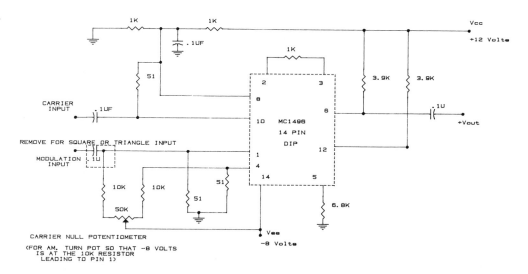

FIGURE 6–5

Because the presence of a carrier at the receiver makes the signal relatively easy to demodulate, a dc level is added to the modulation signal to produce amplitude modulation in common AM radios. Equation 6–4 shows the mathematics of AM modulation for a 100-kHz, 1-V carrier and a 5-kHz, 0.5-V sinusoidal modulation.

$$[1 \cos(2 \times 100,000t)][1 + 0.5 \cos(2 \times 5000t)]$$

$$= 1 \cos(2 \times 100,000t) + 0.5 \cos(2 \times 100,000t) \cos(2 \times 5000t)$$

$$= 1 \cos(2 \times 100,000t) + 0.25 \cos(2 \times 105,000t) \tag{6–4}$$

$$+ 0.25 \cos(2 \times 95,000t)$$

The result of Equation 6–4 is a signal at the carrier frequency (100 kHz) and two smaller signals at the sum and difference frequencies. The amplitude of the modulation signal determines the *modulation index, m*. This index must be less than or equal to 1 (here $m = 0.5$). A plot of Equation 6–4 is shown in Figure 6–3. This waveform can be thought of as the modulation waveform "riding" on the carrier waveform, or modulating the amplitude of the carrier. Note that the 5 kHz signal is carried on both the positive and negative amplitudes of the carrier. A plot of the Fourier analysis of the AM-DSB waveform is shown in Figure 6–4. This is approximately what is seen on a spectrum analyzer for single-frequency AM. Figure 6–5 shows the circuit that you will use for this experiment using the MC1496 balanced modulator/demodulator.

PROCEDURE AND QUESTIONS/PROBLEMS

Procedure

(NOTE: You will use this circuit again in Experiment 7.)

Frequency Doubling

1. Construct the circuit of Figure 6–4, double-checking all connections and all resistor values (this is a very difficult circuit to wire correctly the first time). A single wrong resistor or missed connection will cause incorrect biasing and distort the waveform so that it is unrecognizable (95 percent of the problems in 4 years of use have been due to a misconnection on this circuit). This IC has proven to be rugged unless you reverse the polarity of the power supplies.

2. Connect the same 5 kHz sinusoidal signal (note that inputs of 0.1 to 0.4 V appear to be acceptable, but if there is distortion, reduce the amplitude) to both the carrier and modulation inputs. Observe the output on an oscilloscope and adjust the carrier null potentiometer (see figure) until the output is a 10 kHz sinusoid. Note that this is a very sensitive adjustment because you are making the biasing at both inputs *exactly* the same to get the multiplying effect of the device. Check the input and output frequencies using a frequency counter and verify that you have exact multiplication by 2 in frequency.

3. Change the frequency of the input and verify that you have multiplication at 100 kHz and 500 kHz. Note that there is a decrease in amplitude at the higher frequencies, but the multiplying action continues.

AM/DSB-SC

4. Apply a 100 kHz, 0.1 V peak sinusoid to the carrier input and a 5 kHz, 0.1 V peak sinusoid to the modulation input.

5. Adjust the carrier null potentiometer to obtain a waveform like the one in Figure 6–1. Label the axes of Figure 6–1 to correspond to your waveform. If a spectrum analyzer is available, observe and sketch the output in the frequency domain.

6. Remove the sinusoidal modulation (and the modulation input capacitance) and apply a 0.1 V peak, 5 kHz triangular modulation (ensure that there is *no* dc offset on the triangular modulation prior to making the connection. Adjust the carrier null potentiometer (if required), and label the axes of Figure 6–2 to correspond to your waveform. If a spectrum analyzer is available, observe and sketch the output in the frequency domain.

7. Remove the triangular modulation, and repeat step 3 with a 0.1 V peak, 5 kHz squarewave modulation. Sketch the output after adjusting the carrier null potentiometer.

8. Calculate and sketch the frequency-domain representation of the output of the three modulation waveforms.

AM/DSB

9. Apply a 100 kHz, 0.1 V peak sinusoid to the carrier input and a 5 kHz, 0.05 V peak sinusoid to the modulation input. Turn the carrier null potentiometer so that the -8 V is applied to the 10 k resistor connected to the modulation input. This will provide an AM waveform at the output with a modulation index of about 0.5. Observe the axes and label the waveform of Figure 6–3 to correspond to your waveform. If a spectrum analyzer is available, observe and sketch the frequency-domain representation.

10. Vary the modulation index (amplitude of the modulation waveform) from 0 to 1 and note the variation in the output waveform. Increase the modulation slightly beyond 1 and note the distortion. Explain what the distortion means in the frequency domain.

11. Using square and triangular modulation waveforms, observe the output for various modulation indices.

12. If a spectrum analyzer is available, observe and sketch the frequency-domain representation of the output.

13. Using a 1 MHz carrier and 1 kHz modulation with a modulation index of less than 1, place an AM radio near the output of the 1496. You should be able to hear the tone on the radio if you tune to the vicinity of your 1 MHz carrier (1000 on your radio).

Questions/Problems

1. Mathematically verify that frequency doubling takes place using the following sinusoidal input:

$$v(t) = \sin(2 \times 100{,}000t)$$

2. Mathematically verify DS-SC output using the following inputs (note that 10^6 represents 1 MHz):

$$0.1 \cos(2\pi \times 10^6 t) \quad \text{and} \quad 0.1 \cos(2\pi \times 2000t)$$

3. Mathematically verify an AM/DSB output using the following inputs:

$$0.1 \cos(2\pi \times 10^6 t) \quad \text{and} \quad 0.1[1 + 0.05 \cos(2\pi \times 2000t)]$$

This represents tone modulation of a signal in the middle of the U.S. commercial AM radio band.

4. How could a single-sideband signal be produced from the AM/DSB-SC signal that you generated in the experiment?

5. Use a computer program to plot the equation of Question 1.

6. Use a computer program to plot the equation of Question 2.

7. Use a computer program to plot the equation of Question 3.

7

AM DEMODULATION

INTRODUCTION

In Experiment 6 you constructed an AM modulator and were introduced to the closely related topics of frequency multiplication and suppressed carrier modulation. The purpose of this experiment is to introduce you to demodulation of an AM waveform using a tuned filter, a diode, and a low-pass filter. This simple circuit is the basis for "crystal radio" receivers, the first experiment many of us attempt when we initially become interested in electronics.

After completing this experiment, you will be able to

1. Understand the principles of AM demodulation.
2. Build and tune a simple AM demodulator.
3. Understand tuning of a common AM receiver.

REFERENCES

1. Young, Chapter 5.
2. Roddy and Coolen, Chapter 8.
3. Miller, Chapter 3.
4. Tomasi, Chapter 4.
5. Killen, Chapter 8.
6. Adamson, Chapters 4 and 16.

MATERIALS OR SPECIAL INSTRUMENTATION

Devices: 1N34 (or other Ge) diode

Resistor: 10 k

Capacitors: 150 pF, 0.1 μF, 0.01 μF, 10 to 150 pF (adjustable)

Inductor: 100 μH (self-wound or RF for high Q)

THEORETICAL BACKGROUND

Figure 6–3 of Experiment 6 is a graph showing an AM waveform with a modulation index of 0.5. The desired information is the lower-frequency sine wave modulating the higher-frequency carrier amplitude. The frequencies used for the graph are a 5 kHz modulation and a 100 kHz carrier, but the graph is valid for all AM frequencies by changing the time axis. Of course, a voice signal is composed of many different frequencies and would show up on an oscilloscope as a constantly changing carrier amplitude, but the same AM principles apply.

Demodulation of this AM signal requires recovery of the lower-frequency sine wave. With a bit of thought, it is easy to see that the result of half-wave rectifying the waveform in Figure 6–3 is the waveform shown in Figure 7–1.

If the high-frequency signal could be eliminated (by a low-pass filter), the desired signal would be left. Because the low- and high-frequency signals are far apart in frequency (a commercial-band AM radio has a 0 to 5 kHz frequency range with a carrier

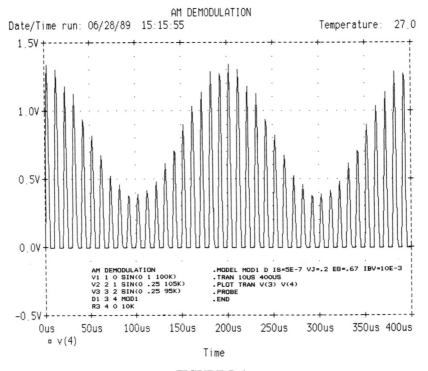

FIGURE 7–1

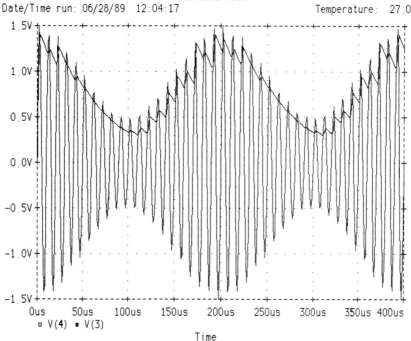

AM DEMODULATION
Date/Time run: 06/28/89 12:04:17   Temperature: 27.0

FIGURE 7–2

frequency of 0.54 MHz to 1.6 MHz), a simple (single-pole) low-pass filter can be used to effectively eliminate the high-frequency signal to produce the demodulated, original audio frequency signal shown in Figure 7–2. Note in Figure 7–2 that the demodulated signal is not a perfect representation of the modulating signal because a 5 kHz signal and a 100 kHz carrier were used (for a presentable drawing). Thus, the two signals are only a factor of 20 apart in frequency, and the higher frequency signal is not totally eliminated. Your AM signal uses a 1 kHz signal and a 1 MHz carrier, which are a factor of 1000 apart in frequency, so that the 1-pole filter used provides good elimination of the 1 MHz carrier.

Figure 7–3 shows the circuit to be used in demodulating the AM signal for this experiment. The output from the 1496 balanced modulator in experiment 6 will be capacitively coupled (to eliminate the 1496's bias) to the filter circuit. This high-Q

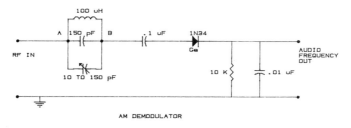

AM DEMODULATOR

FIGURE 7–3

resonant circuit provides a voltage gain (note that this is a passive circuit, so there can be no power gain) at resonance that is sufficient to enable the Ge diode to rectify the AM input. In an AM radio, a similar resonant filter is tuned to the desired AM frequency to select the desired station. Note that the resonant circuit is not necessary here with only a single-frequency input if the input voltage is large enough to enable the diode to rectify the AM signal.

To convert the 1496 modulator circuit into an AM transmitter would require a power amplifier at the output of the 1496 and a matching device to feed an AM antenna. To convert the simple detector used in this experiment to receive electromagnetic waves would require an AM antenna (generally ferrite), a tuner (similar to the resonant circuit used), and amplifiers to boost the RF signal to sufficient strength for the diode to rectify it without eliminating part of the audio signal.

PROCEDURE AND QUESTIONS/PROBLEMS

Procedure

1. Construct the circuit of Figure 7–3.
2. Using a 1 kHz modulating frequency and a 1 MHz carrier with the 1496 from Experiment 6, produce an AM signal with the highest possible undistorted output at pin 6 of Figure 6–4.
3. Capacitively couple the output of pin 6 of the 1496 to the input of the circuit of Figure 7–3 using a 0.1 μF capacitor (as shown in Figure 6–4).
4. Measure and record the voltages at points A and B on Figure 7–3. Comment on the voltage difference at resonance.
5. Observe the circuit output with and without the 0.01 μF filter capacitor and verify the operation of this AM detector/demodulator.

Questions/Problems

1. Determine the capacitance value required in the resonant circuit to tune to 1 MHz.
2. Why is a Ge diode used instead of a Si diode?
3. How could you convert the AM modulator/demodulator combination into a system to transmit and receive voice signals using your house wiring? (Note that you need a very sharp 60 Hz band-stop filter in the receiver.)

8

A TUNED-INPUT/TUNED-OUTPUT RF AMPLIFIER

INTRODUCTION

RF amplifiers are required in transmitters and receivers and are defined to be amplifiers used at "high" frequencies. Here, the high frequency to be amplified is 10.7 MHz, which is the standard intermediate frequency (IF) used in commercial FM radios. Use of RF amplifiers usually requires tuned circuits to ensure that only a limited bandwidth (generally the BW of the desired signal) is passed through the amplifier. Generally, band-pass filter circuits are used at both the input and output of the circuit. These can be tuned to the same frequency or to two slightly different frequencies (stagger-tuning). In this experiment you will construct and measure the characteristics of a 10.7-MHz RF amplifier and observe the effect of tuning two band-pass filters in the same circuit.

After completing this experiment you will be able to

1. Understand the principles of tuning two band-pass filters for a desired bandwidth.
2. Understand the advantages of a cascode amplifier configuration.
3. Construct an RF amplifier.

REFERENCES

1. Young, Chapters 1 and 7.
2. Roddy and Coolen, Chapter 5.
3. Miller, Chapter 3.
4. Tomasi, Chapter 4.
5. Killen, Chapter 6.
6. Adamson, Chapter 14.

MATERIALS OR SPECIAL INSTRUMENTATION

10.7-MHz signal generator

Spectrum analyzer (if available)

Devices: CA3028A or CA3053 or ECG724, differential amplifier

Resistors: 50 Ω, 1 k, 2 k

Capacitors: 39 pF, 36 pF, 470 pF, two 8–35 pF, three 0.001 μF

Inductors: two 3–5 μH RF (if you wind your own, make them about 4 μH each)

THEORETICAL BACKGROUND

An analysis of Figure 8–1, an IC RF amplifier, will introduce some interesting theoretical aspects of RF amplification.

Pins 1 and 8 are shorted together to completely eliminate Q_1 from the circuit. Q_1 is half of the differential amplifier pair and is used when a differential amplifier is constructed. The input is provided to the base of Q_3, which is a common-emitter (CE) amplifier. Q_2 is a common-base (CB) amplifier, which provides the load at the collector of Q_3 and makes this a classic cascode amplifier (in an integrated circuit). The voltage gain of the cascode CE amplifier is about 1, and the current gain of the cascode CB amplifier is also about 1. The cascode configuration is frequently used in RF amplifier applications because of its good gain (the current gain is due to the CE portion, and the voltage gain is due to the CB portion) and excellent isolation between the input and the

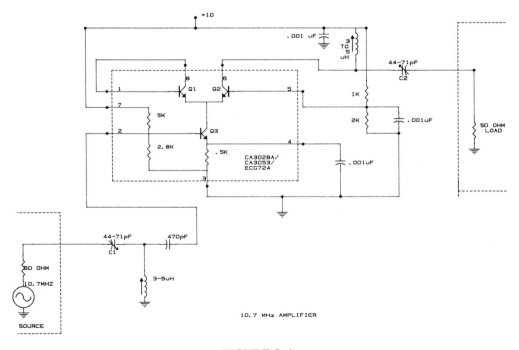

FIGURE 8–1

56

output. There is virtually no Miller-effect feedback from output back to the input, and neutralization is usually not required.

The advantage of the IC over discrete transistors is the matching of the transistors and the accurate internal biasing resistors. Analysis of the biasing of the circuit is identical to using discrete devices. The 0.001 μF capacitor from pin 4 to ground is just an emitter bypass resistor. Similarly, the 1 k and 2 k resistors (and 0.001 μF capacitor) attached to pin 5 provide biasing (and the capacitor provides high-frequency bypass) for Q_2. Q_3 is automatically (internally) biased with pin 7 connected to the power supply.

Since all the biasing is straightforward, the input and output circuits are the critical portions of the amplifier. Because the input and output of the IC have a dc level, they must both be ac coupled. The 470 pF capacitor provides ac coupling of the input (pin 2). The ac coupling to the load is provided by the tuning capacitors (pin 6).

Design of an RF amplifier using this IC is basically design of the band-pass filters and the RF transformers (or other methods of interfacing the input and output of the amplifier). The circuit shown in Figure 8–1 uses 50 Ω input and load impedances. These are common in RF work, but other impedances could easily be used. The band-pass filters at the input and output must use inductors and capacitors that allow tuning slightly above and below 10.7 MHz. A high-Q coil (and therefore high-Q circuit) is normally required if narrow bandwidth is desired. The input and output coupling of the circuit generally use RF transformers to couple the small (50 Ω) input and output impedances into the circuit without affecting the Q's of the tuned circuits; however, direct coupling is used here. If both variable inductors are set to the center of their

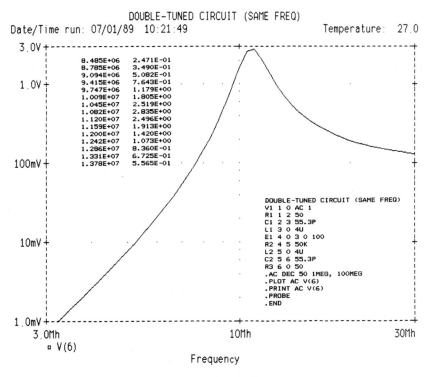

FIGURE 8–2

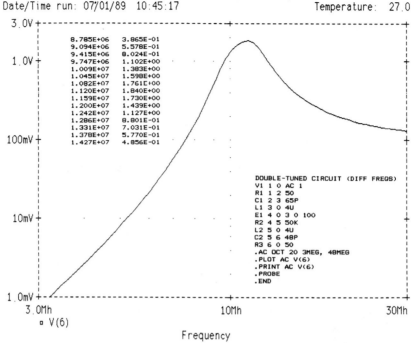

8.785E+06	3.865E-01
9.094E+06	5.578E-01
9.415E+06	8.024E-01
9.747E+06	1.102E+00
1.009E+07	1.383E+00
1.045E+07	1.598E+00
1.082E+07	1.761E+00
1.120E+07	1.840E+00
1.159E+07	1.730E+00
1.200E+07	1.439E+00
1.242E+07	1.127E+00
1.286E+07	8.801E-01
1.331E+07	7.031E-01
1.378E+07	5.770E-01
1.427E+07	4.856E-01

```
DOUBLE-TUNED CIRCUIT (DIFF FREQS)
V1  1  0  AC  1
R1  1  2  50
C1  2  3  65P
L1  3  0  4U
E1  4  0  3  0  100
R2  4  5  50K
L2  5  0  4U
C2  5  6  48P
R3  6  0  50
.AC OCT 20 3MEG, 48MEG
.PLOT AC V(6)
.PRINT AC V(6)
.PROBE
.END
```

Frequency

FIGURE 8–3

range (4 μH), and the variable capacitors are set to about the center of their range (55.3 pF), both band-pass filters are tuned to 10.7 MHz. Thus, using the inductor and capacitor values shown will provide a wide tuning range for the circuit. Both filters being tuned to 10.7 MHz will provide the minimum bandwidth for the system and provide maximum gain (as is shown in Figures 8–2 and 8–3).

Figure 8–2 shows the predicted frequency response with both filters tuned to 10.7 MHz ($L = 4$ μH, $C = 55.3$ pF). Note that the gain shown in Figures 8–2 and 8–3 does not correspond to the actual circuit voltage gain, since it is just a model to illustrate the tuning. From Figure 8–2, the bandwidth of the amplifier is 1.27 MHz as calculated from the figure and the attached printout. Coupling of the input and output using an RF transformer would allow a higher Q for the filters and provide a narrower amplifier bandwidth.

Figure 8–3 shows the frequency response of the circuit using stagger-tuning to widen the bandwidth of the circuit ($L = 4$ μH, $C_1 = 65$ pF, $C_2 = 48$ pF). This tunes the input circuit to 9.87 MHz and the output circuit to 11.49 MHz. The wider bandwidth of the circuit can be seen by comparing Figure 8–3 and Figure 8–2 or by looking at the two associated printouts. The maximum relative gain of the stagger-tuned circuit compared with identical tuning is 1.9 versus 2.9, so there is very little loss of gain (a little over 3 dB). The bandwidth of the stagger-tuned circuit has now been increased to about 2.24 MHz with a relatively flat gain response from about 10.5 MHz to 11.7 MHz (only about 1.5 dB loss over this range). Stagger-tuned circuits can be used with more than two

58

stages and can be tuned to exhibit Butterworth, Chebyshev, or other desired filter characteristics.

PROCEDURE AND QUESTIONS/PROBLEMS

Procedure

1. Construct the circuit of Figure 8–1 with the following notes.
   a. If you wind your own inductors, make them about 4 μH.
   b. You can tune the circuit using a single variable capacitor or a fixed capacitor in parallel with a small variable capacitor. Be careful of the effect you have on the circuit while tuning. Change the capacitance (or inductance), then move away from the circuit while taking measurements.
   c. Use short leads, and do not use long wires unless they are shielded, or you may turn this amplifier into an oscillator.
2. Measure the inductance and capacitance values and tune both band-pass circuits to 10.7 MHz. Your inductance and capacitance values may differ slightly from those shown, but a relatively high Q inductor is necessary.
3. Measure the voltage gain of the amplifier at 10.7 MHz. Calculate the power gain of the amplifier at this frequency. Comment on the effectiveness of your amplifier.
4. Measure the 3 dB bandwidth of your amplifier with both band-pass filters tuned to 10.7 MHz. Use the spectrum analyzer, if available, or manually sweep the input frequency from the center frequency (10.7 MHz) to the lower frequency that produces an output that is 0.707 of the maximum voltage and then to the higher frequency that produces 0.707 of the maximum voltage. Comment on your measured bandwidth versus the bandwidth calculated from Figure 8–2.
5. Measure the voltage gain of your amplifier using a 220 Ω load resistor. Calculate the power gain, and comment on the difference in gains between a 220 Ω load and a 50 Ω load.
6. Tune the input band-pass filter to 9.87 MHz and the output band-pass filter to 11.49 MHz (calculate the L's and C's required). Measure the new bandwidth. Also, measure the maximum gain of the circuit in this configuration. Comment on the results.

Questions/Problems

1. Calculate the range of tunable frequencies available with the band-pass filters shown in Figure 8–1.
2. Design a 50 MHz amplifier using the same configuration by stating the changes that would have to be made to Figure 8–1 to convert to the higher frequency.
3. Discuss additional considerations that would have to be made for a higher frequency amplifier (decoupling, lead length, and the like).
4. What is the approximate input resistance of the 10.7-MHz RF amplifier?
5. Design a stagger-tuned filter for the 50 MHz amplifier of Question 2 using a computer program.

9

TIME-DIVISION MULTIPLEXING

INTRODUCTION

Time-division multiplexing (TDM) is widely used in communications for transmission of digital signals. Fiber-optic cables use TDM for transmission of multiple signals, so the extensive use of fiber optics has increased the use of TDM worldwide. In this experiment, you will time-division multiplex four different signals using a TTL counter and a CMOS analog multiplexer and observe the output waveform. This will provide visual understanding of actual TDM concepts and methods.

After completing this experiment you will be able to

1. Understand time-division multiplexing and demultiplexing.
2. Understand the operation of a 4-channel analog multiplexer.
3. Build and use a time-division multiplexer.

REFERENCES

1. Young, Chapter 11.
2. Roddy and Coolen, Chapter 11.
3. Miller, Chapter 9.
4. Tomasi, (*Advanced*), Chapter 5.
5. Killen, Chapter 11.
6. Adamson, Chapter 8.

MATERIALS OR SPECIAL INSTRUMENTATION

Devices: 555 timer, 7493 TTL counter, 4052 dual 4-channel analog multiplexer

Resistors: five 1 k, 4.7 k, 5 k potentiometer

Capacitor: 0.1 μF

THEORETICAL BACKGROUND

TDM allows many signals to be placed on a single high-bandwidth (therefore, high bit rate) channel for efficiency of transmission. For instance, a single-mode fiber-optic cable containing seventy-five 200 Mbit/s fiber-optic waveguides can transmit up to $75 \times 200 \times 10^6 = 15$ Gbits/s of information (this is within the current state of the art). A telephone conversation is bandwidth limited to 3400 Hz (not exactly high fidelity) and sampled at 8000 samples/second. If each sample is pulse-code modulated in 8 bits, then 64,000 bits/s must be transmitted for each telephone conversation (this is the commercial rate). This means that, using TDM, $15 \times 10^9/64,000 = 234,000$ separate telephone conversations could be carried over a single cable, and $234,000/75 = 3125$ conversations could be carried over a single fiber.

Since TDM is a form of modulation, you must consider your ability to demodulate the signal at the receiver. Nyquist's sampling theorem must always be the fundamental consideration for TDM to be able to demodulate the signal after transmission. Nyquist proved that a signal must be sampled at a rate that is at least twice the highest frequency of the signal. Therefore, a 100 Hz signal must be sampled at least 200 times per second, a 500 Hz signal 1000 times per second, and so forth. Note that this is a theoretical minimum sampling rate. The telephone companies (true models of electronic efficiency) sample the band-limited, 3400 Hz voice signals at 8000 bits/s, not at the theoretical minimum of 6800 bits/s. So one must actually sample faster than the Nyquist rate for any real signal.

TDM requires multiplexing of different signals sequentially. There are many ways of doing this electronically. In this experiment, you will use an easily understandable method that combines several ICs that you have used before (555 timer and a 4-bit TTL counter) with an analog multiplexer. An analysis of Figure 9–1, which you will construct, will provide a clear understanding of TDM.

The 555 timer is used as a 50 percent duty cycle oscillator that provides a squarewave clock signal. The 555 in the circuit can be replaced by a squarewave signal generator operating at the desired TDM rate if desired. The 555 simulates the clock signal that would generally operate at higher than the 1 kHz rate used here and may be computer generated for control of the multiplexing rate. The 1 kHz clock signal from the output (pin 3) of the 555 is fed directly into the 7493, 4-bit, binary TTL counter. You are using the 4052 dual 4-channel multiplexer (and using only one set of four channels) and need to use only 2 bits of the 7493 to repeatedly count 00, 01, 10, 11, 00, 01, and so on. If you used the 4051 analog multiplexer, you could multiplex eight channels using the 7493 by using three of its bits for repetitive counting.

The output of the 7493 (bits 0, pin 9, and bit 1, pin 8, for your purposes) is fed directly into inputs A and B, respectively, of the 4052. Input A, pin 10, and input B, pin

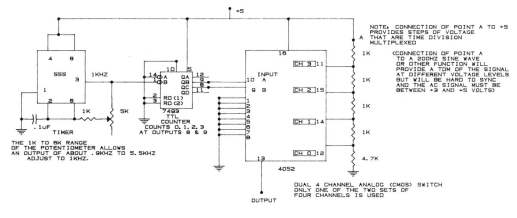

NOTE: CONNECTION OF POINT A TO +5
PROVIDES STEPS OF VOLTAGE
THAT ARE TIME DIVISION
MULTIPLEXED

(CONNECTION OF POINT A
TO A 200HZ SINE WAVE
OR OTHER FUNCTION WILL
PROVIDE A TDM OF THE SIGNAL
AT DIFFERENT VOLTAGE LEVELS
BUT WILL BE HARD TO SYNC
AND THE AC SIGNAL MUST BE
BETWEEN +3 AND +5 VOLTS)

DUAL 4 CHANNEL ANALOG (CMOS) SWITCH
ONLY ONE OF THE TWO SETS OF
FOUR CHANNELS IS USED

TIME DIVISION MULTIPLEXING

FIGURE 9–1

9, of the 4052 control which of the four input channels (0, 1, 2, or 3) is fed to the output (pin 13). Thus, the 4052 output is directly controlled by whether inputs A and B are 00, 01, 10, or 11. Since the 7493 provides these signals as it repeatedly counts up, the channels are continuously sequentially selected, and the result is a TDM signal.

The simplest TDM signal to observe is different voltage levels at the input of each of the four channels. The resistor divider network shown in Figure 9–1 provides a different dc voltage at each input channel. The output on pin 13 of the 4052 is easily observed to be TDM. Different signals may be provided at each of the four input channels of the 4052 and a more realistic TDM signal produced; however, the waveforms of each of the four channels will not be readily observable due to the difficulty of synchronizing individual TDM signals with the oscilloscope.

PROCEDURE AND QUESTIONS/PROBLEMS

Procedure

1. Construct the circuit of Figure 9–1.
2. Ensure that the 555 output (or the output of a signal generator that you may use to replace it) is a squarewave at TTL levels. 1 kHz was used in the figure for convenience only. Much higher counting rates are possible with the 7493/4052 combination.
3. Observe and sketch the oscilloscope output of the 4052 (pin 13).
4. Connect point A (at the upper right of the figure) to a sinusoidal source at less than 500 Hz with a 3 V dc level and a peak of 5 V. Sketch, if possible, and comment on the resulting waveform. Why must the signal source be less than 500 Hz with a 1 kHz clock rate?
5. Connect a 300 Hz square wave to one of the input channels of the 4052 and a 300 Hz sine wave to another input channel, ensuring that the levels are between 3 and 5 V. Ground the other two channels. Sketch and comment on the resulting waveform.

QUESTIONS/PROBLEMS

1. Describe time-division multiplexing in your own words.
2. Design an 8-channel TDM system using a 4051.
3. Assume that you have eight analog signals. You use your 8-channel multiplexer design to send one of the eight signals each 0.0125 ms. What is the maximum frequency of the signals? What is the clock rate required?

10

FREQUENCY MODULATION

INTRODUCTION

The use of a voltage-controlled oscillator (VCO) with a varactor diode is a common, and easily understood, method of generating frequency modulation (FM). In this experiment you will construct an FM modulator, observe single-frequency FM in the time and frequency domains, and learn the operation of a VCO and a varactor diode. In the next experiment, you will construct a single-chip FM receiver.

After completing this experiment you will be able to

1. Understand the principles of a VCO.
2. Understand varactor diode control of a VCO.
3. Build and tune a simple FM modulator.
4. Recognize a single-frequency FM waveform and its associated frequency spectrum.

REFERENCES

1. Young, Chapter 9.
2. Roddy and Coolen, Chapter 10.
3. Miller, Chapter 5.
4. Tomasi, Chapter 7.
5. Killen, Chapter 9.
6. Adamson, Chapters 6 and 15.

MATERIALS OR SPECIAL INSTRUMENTATION

Spectrum analyzer (if available)

Graph paper

Devices: MC1648 VCO, MV 2112 varactor diode [Data: (voltages are reverse bias) 56 pF at 3 V; 22 pF at 30 V]

Capacitors: three 0.1 μF, 0.01 μF

Inductor: 100 μH (high Q for best results)

THEORETICAL BACKGROUND

In amplitude modulation the carrier amplitude varies directly with the modulation amplitude. In frequency modulation the carrier frequency varies directly with the modulation amplitude with no variation in the amplitude of the carrier. This condition can be represented by Equation 10–1.

$$f(t)_{FM} = A \cos[6.283 f_c t + m \cos(6.283 f_m t)] \tag{10–1}$$

where

f_c = carrier frequency

m = modulation index

f_m = frequency modulation

This is not as simple a mathematical expression as that for AM, but you can see from Equation 10–1 that (dependent on the value of the FM modulation index, m) the value in the brackets (the total phase of the waveform) will vary. The carrier frequency is normally much higher than the modulation frequency, so that there is a rapid change due to the carrier component of the equation and a slower change due to the lower frequency modulation component. Thus, the modulation "slowly" (relative to the very fast carrier) varies the frequency of the overall signal. Note that the amplitude of the overall signal stays the same.

Figure 10–1 shows an FM waveform with a 100-kHz carrier frequency, a modulation index of 10 (rather high so that the graph clearly shows the FM waveform), and a single frequency modulation at 5 kHz. Note that because of the large modulation index, a small amount of distortion is evident in Figure 10–1 as a minor variation in the amplitude of the signal. This distortion is not allowable in commercial FM systems. However, amplitude distortion like that shown in Figure 10–1 does occur due to transmission phenomena, so FM receivers generally have a limiter. A limiter clips the peak amplitude of the FM signal to a specified level, so there is no variation prior to FM detection.

Figure 10–2 shows the frequency response of the FM signal in Figure 10–1. This is similar to what is seen on a spectrum analyzer. Note that for single-frequency modulation at 5 kHz, Figure 10–2 shows many (symmetry is always present) sidebands at 5 kHz intervals around 100 kHz. The number and amplitude of the sidebands can be predicted using a chart of Bessel functions and the modulation index of the signal. The bandwidth of the FM signal in Figure 10–2 is about ±80 kHz, although you can predict that small sidebands still exist beyond these frequencies.

A method of determining the modulation index, m, of a single-frequency FM signal is to determine the maximum frequency deviation of the carrier ($\pm f_d$) and use

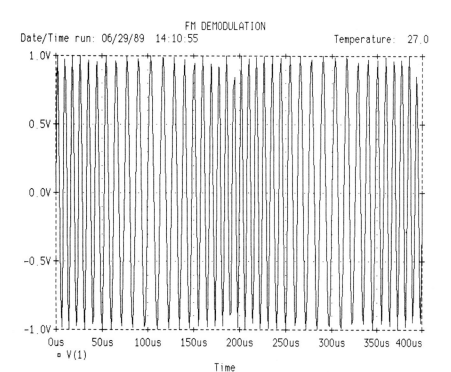

FIGURE 10–1

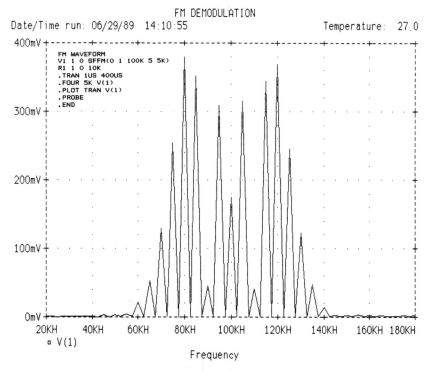

FIGURE 10–2

Equation 10–2 or determine the k_0 (a constant of the VCO) in hertz/volt (or kilohertz/volt) and use Equation 10–3. Note that the maximum frequency deviation (f_d) in Equation 10–2 is the same as $k_0 V_m$ in Equation 10–3.

$$m = \frac{f_d}{f_m} \qquad (10\text{–}2)$$

where
f_d = maximum frequency deviation

f_m = modulation frequency

$$m = \frac{(k_0 V_m)}{f_m} \qquad (10\text{–}3)$$

where
k_0 = kilohertz/volt

V_m = the peak modulation voltage

f_m = the modulation frequency

    Carson's rule for the bandwidth of an FM signal, Equation 10–4, predicts a bandwidth of 110 kHz for the signal shown in Figures 10–1 and 10–2.

$$BW = 2f_m(1 + m) \qquad (10\text{–}4)$$

    The circuit to be used for this experiment is shown in Figure 10–3. The basis of the circuit is an oscillator (the output is a square wave; however, a sine wave can be obtained, see specifications) whose frequency is controlled by a tuned (resonant *LC*) circuit at the input. Early VCOs used a single-transistor oscillator; however, the MC1648 used in this experiment can provide an output up to 200 MHz with proper compensation. The input resonant circuit is a tank (parallel *LC*) circuit composed of the 100 μH inductor in parallel with the input capacitance of the MC1648 (about 6 pF) and the MV 2112 varactor diode. The varactor is in parallel with the inductor at the resonant frequency (in the megahertz range) because the 0.01 μF capacitor at the front end of the circuit is a short circuit at the resonant frequency. For instance, at a resonant frequency of 2.5 MHz (which is the approximate range you will initially use), the impedance of the 0.01 μF capacitor is 6.37 Ω, which is very small compared with the input impedance of the circuit to the audio frequency signals (the impedance at 5 kHz is 3.2 Ω) used for modulation.

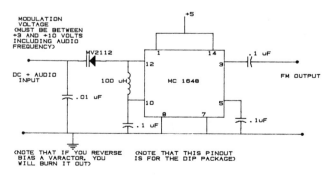

FIGURE 10–3

The MV 2112 varactor diode is a voltage-variable capacitor and is the key to the circuit. The diode must always be reverse biased or it will burn out. It operates in the range of 3 to 30 V reverse bias. When it is reverse biased, it acts as a pure capacitance that varies with the voltage across the diode. The varactor is about 56 pF at 3 V reverse bias and about 22 pF at 30 V reverse bias. The variation with voltage is nonlinear, but over a small range can be approximated as a linear function.

PROCEDURE AND QUESTIONS/PROBLEMS

Procedure

VCO
1. Measure and record the inductance of your 100 μH inductor.
2. Construct the circuit of Figure 10–3.
3. Calculate the range of resonance (and, therefore, frequency of oscillation) for a 3 to 15 V reverse bias on the varactor. (Assume the capacitance varies linearly for this approximate calculation.) This will give you a good estimate of the general range of the oscillation.
4. Using a dc input of 3 to 15 V, measure, record, and plot the output frequency (vertical axis), using a frequency counter, versus the input (dc) voltage. (This should be about 1.8 to 2.5 MHz). This graph provides the characteristic of the VCO in volts versus hertz. A VCO is generally specified in terms of its variation of frequency with voltage (often called k_0 and given in kilohertz/volt). This is simply the inverse of the slope of the graph you have plotted. Calculate and record the VCO constant (k_0) of your VCO in kilohertz/volt at 6 V. You will see this constant again when discussing and calculating phase-locked loop characteristics in later experiments.

Frequency Modulation
5. Provide a 0.1 Hz (a very slow change that is observable to the human eye), 1 V peak squarewave input with a 6 V dc value to the VCO (the squarewave should vary from 5 to 7 V). Observe the frequency counter for a few minutes and explain the output.
6. Change the input frequency to a 1 V peak, 5 kHz sine wave (still with a 6 V dc level for reverse bias of the varactor). Sketch and explain the output on the oscilloscope and the value shown on the frequency counter.
7. Calculate and record the modulation index of your FM signal (Equation 10–3).
8. Measure the maximum frequency variation of your FM signal output on the spectrum analyzer (if available). From Equation 10–2 determine the modulation index of your signal. Discuss the difference between this measurement and the calculation of part 3.
9. Using the spectrum analyzer on the FM signal with the same 5 kHz sine wave (1 V peak with 6 V of dc bias), adjust the amplitude of the sinusoid (the dc stays at 6 V) until the carrier frequency is first nulled out (0 amplitude) on the spectrum analyzer. At this point (if you are using the first null), the modulation index, m, is 2.4, as you can verify by looking at a table of Bessel functions. At this point, use the DMM (on ac) to measure and record the modulation amplitude (remember, you are generally measuring rms with the DMM and must multiply by 1.414 to obtain the peak value). Knowing m, V_m, and f_m, you can use Equation 10–3 to calculate the deviation

constant of your VCO. Compare this measured value with the value you calculated from the graph in part 4. Comment on differences between the two values.

Questions/Problems

1. For an 80 nH coil (a few turns of #22 wire with no ferrite core), determine the capacitance required for the input tank circuit to be resonant at 100 MHz (the approximate center of the commercial FM band).
2. For your MV 2112 varactor, determine the approximate dc bias for a center frequency of 100 MHz. (Project your measurements to the higher range.)
3. For a ±75-kHz frequency deviation (a 150 kHz bandwidth is standard for the commercial FM radio band) determine the modulation index of a 15 kHz modulation signal (standard FM radio audio bandwidth) from Equation 10–2.
4. For the FM modulator you designed in the first three questions, using the k_0 that you calculated in lab for your VCO and Equation 10–3, calculate the maximum modulation voltage required at the input. [You know k_0, m (Question 3), and f_m (Question 3)]. Note that k_0 will be different at the higher bias you would need for this frequency range.

11

SINGLE-CHIP FM RADIO CIRCUIT

INTRODUCTION

A superheterodyne FM radio receiver is time consuming and difficult to construct when each of the components must be separately designed and built. The TDA7000 single-chip receiver allows a rapid and interesting look at the theory and components used in an FM radio receiver. The purpose of this experiment is to build and test an FM radio receiver. In the process, you will tune a VCO, make a simple one-quarter-wavelength FM antenna, build a band-pass input filter, build an audio amplifier (if necessary), and learn the operating principles of the TDA7000 single-chip FM radio IC.

   After completing this experiment, you will be able to

1. Construct a single-chip FM radio using the TDA7000.
2. Understand the principles of tuning and demodulation used in the TDA7000.
3. Better understand many of the principles you applied individually and how they relate to a system.

REFERENCES

1. Young, Chapters 7 and 9.
2. Roddy and Coolen, Chapters 7 and 10.
3. Miller, Chapter 6.
4. Tomasi, Chapter 8.
5. Killen, Chapter 10.
6. Adamson, Chapters 6 and 16.

MATERIALS OR SPECIAL INSTRUMENTATION

FM Receiver

Device: TDA7000 single-chip FM radio

Resistor: 22 k

Capacitors: (one each)

(in pF) 10 to 126 variable, 27, 39, 47, 56, 150, 180, 220, 330, 330

(in nF) 1.8, 2.2, 3.3, 3.3, 10, 10, 22, 100, 150

Inductors: 130 nH, 56 nH (You can easily make these; see procedure.)

Workbench Audio Amplifier

Devices: LM386 low-voltage power amplifier, 8 Ω speaker

Resistors: two 4.7 k, 5 k potentiometer, 10 k, 18 k

Capacitors: 0.01 μF, 0.1 μF, 1 μF, 10 μF, 220 μf

THEORETICAL BACKGROUND

An audio amplifier is required at the output of the FM receiver. A simple, single-chip audio amplifier is shown in Figure 11–1, if required. The gain of this amplifier is set to 200 by the capacitor between pins 8 and 1 of the IC but can be set to any value between 20 and 200 that is desired (see specifications in the Appendix).

A block diagram of the TDA7000 single-chip FM radio circuit is shown in Figure 11–2*. A discussion of the function of each of the components of the system will provide an overview of the general theory of FM demodulation.

General

Most AM and FM radio receivers use an intermediate frequency (IF). This IF is usually 455 kHz for AM and 10.7 MHz for FM commercial receivers. These frequencies were

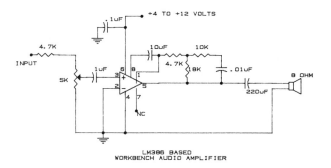

FIGURE 11–1

---

* ARCHER®, Technical Data, *TDA7000, A Complete F.M. Radio on a Chip* (Fort Worth: Tandy Corporation, copyright 1987).

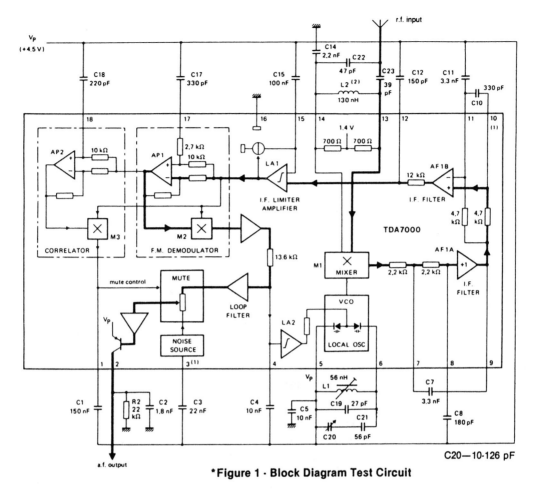

*Figure 1 - Block Diagram Test Circuit

FIGURE 11-2

originally selected due to image frequency (and other) considerations but are used now because of the large number of readily available components made specifically for use at these frequencies. The TDA7000 uses heterodyning (mixing down to an IF) principles, but because of the difficulty in working with (amplifying and filtering) 10.7 MHz (or any high frequency) on an IC, it uses an IF of 70 kHz. Many of the external capacitors used (12 of the chip's pins have capacitors directly to ground) provide filtering when combined with the internal resistors and active devices.

RF Input and Mixer

After the RF signal is received at the antenna, it is passed through a band-pass filter and fed into the mixer through pin 13. Note that the two 700 Ω resistors shown internally between pins 14 and 15 are actually 700 kΩ. The capacitor labeled C_{14} is a bypass capacitor at RF frequencies, and the LC band-pass filter consists of C_{22}, C_{23}, and L_2. Figure 11-3 shows the characteristics of this filter assuming a 37 Ω input impedance

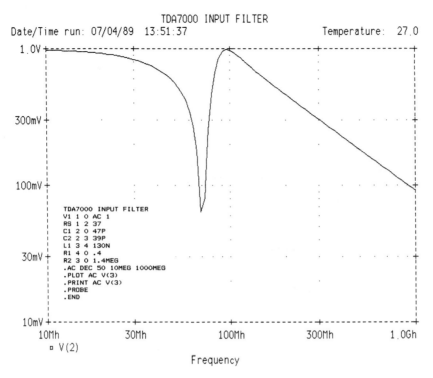

FIGURE 11–3

(the approximate input impedance of the one-quarter-wavelength whip antenna that you will attach to the receiver). Note that the filter has maximum gain throughout the FM radio range (88–108 MHz) and attenuates frequencies outside this range. This reduces the noise input from the antenna into the receiver. The mixer is much like the single-transistor mixer or the 1496 multiplier circuits that you used previously and provides sum and difference frequency outputs.

To use a 70-kHz IF frequency, the other input to the mixer (the VCO) must be 70 kHz above (or below, of course, but 70 kHz above is used here) the desired radio channel frequency. For instance, if you are tuning to 100 MHz, the other input to the mixer must be at 100.070 MHz. The tuning is quite sensitive, but the frequency-locked loop internal to the IC keeps the VCO input to the mixer locked onto the station once you have found a large-enough signal for reception.

VCO

Because the tuning of the VCO is critical, the tank circuit between pins 5 and 6 must be adjustable throughout the 88 to 108 MHz range, with good sensitivity. The VCO and tank circuit input used here is similar to the 1648 VCO that you previously built except for the integrator marked LA_2, internal to the chip, near pin 4. LA_2 is part of the frequency-locked loop that keeps the VCO locked onto a frequency once a strong-enough signal is received, unless it is manually tuned past that frequency. The tuning of the input tank circuit to the VCO determines the frequency of oscillation of the VCO.

74

Thus, for a 70 kHz IF, the VCO must tune in the range of 88.070 to 108.070 MHz. The 56 nH RF coil as well as the capacitor labeled C_{20} are shown to be tunable. A quick calculation shows that the range of tunable frequencies (assuming 56 nH) due to capacitor C_{20} is 82.9 MHz to 113 MHz, as shown in Equations 11–1 and 11–2.

$$f_{low} = \frac{1}{6.283 \sqrt{56 \times 10^{-9} \times 65.8 \times 10^{-12}}} = 82.9 \text{ MHz} \qquad \textbf{(11–1)}$$

$$f_{high} = \frac{1}{6.283 \sqrt{56 \times 10^{-9} \times 35.5 \times 10^{-12}}} = 113 \text{ MHz} \qquad \textbf{(11–2)}$$

The 10 nF capacitor from pin 5 to ground is a virtual short (0.159 Ω) at RF frequencies, as is shown in Equation 11–3.

$$Z_{C5} = \frac{1}{6.283 \times 10^{8} \times 10 \times 10^{-9}} = 0.159 \text{ Ω} \qquad \textbf{(11–3)}$$

Other External Resistors and Capacitors

The general function of the other external elements, beginning at the bottom left of Figure 11–2, is as follows:

C_1 (150 nF) is an RF bypass that determines the time constant for the muting-control circuit (the mute control is active here and eliminates scratchy background interference in the audio signal due to higher frequency interference in the carrier).

The R_2 and C_2 combination (22 kΩ/1.8 nF) is a low-pass filter to eliminate frequencies above the audio range. Equation 11–4 shows the 3 dB cutoff frequency of this filter to be 4 kHz. Thus, this is not a high-fi system.

$$f_{3dB} = \frac{1}{6.283 \times 22 \times 10^{3} \times 1.8 \times 10^{-9}} = 4 \text{ kHz} \qquad \textbf{(11–4)}$$

C_3 is an RF bypass.

C_4 is an RF bypass and provides the low-pass filter required in the frequency-locked loop. The 10 nF combined with the internal 13.6 k resistor forms a LP filter with a 3 dB frequency of 1170 Hz. This type of filter is required in phase- and frequency-locked loops, as will be seen in the phase-locked-loop experiments.

C_7, C_{10}, C_{11}, and internal 2.2 k and 4.7 k resistors along with the internal amplifiers labeled AF1A and AF1B form an active band-pass IF filter that allows only the 70 kHz IF frequency through to the IF amplifier.

C_{12} limits the amplitude of the IF signal. This eliminates amplitude modulation on the FM signal that would cause distortion when demodulation occurs.

C_{15} is large (0.1 uF) and provides a source of charge and hence current to LA$_1$ (IF limiter amplifier).

C_{17} combined with the internal 2.7 k and 10 k resistors provides a phase shift at the input to AP$_1$. The phase-shifted 70 kHz input to AP$_1$ is compared with the original waveform, and the resulting output is mixed (multiplied) by the original waveform in mixer M$_2$, yielding the audio frequency output.

C_{18} is part of the correlator circuit that generates a signal that represents the noise in the IF signal. The output of the correlator controls the muting, which eliminates the staticlike noise from a weak station.

PROCEDURE AND QUESTIONS/PROBLEMS

Procedure

1. Construct the circuit of Figure 11–2. Some hints and notes are provided.

    a. If the 130 nH and 56 nH RF coils are not available, use the following formula (Equation 2–1) with #22 (or so) coated wire:

$$L(\text{solenoid}) = \frac{(N^2 \times R^2)}{(9R + 10X)} \quad \text{(microhenries)}$$

where
                        N = the number of turns

                                  R = the radius of the coil in inches

                                  X = the length of the coil in inches

For instance, with a 1/4-in.-diameter dowel rod, $R = 0.125$ in. If a length X of 1/2 in. is assumed, for $L = 0.13$ uH (130 nH), N is approximately 7 turns, as is shown in Equation 11–5.

$$N = \frac{[L(9R + 10X)]^{1/2}}{R} = 7.14 \tag{11-5}$$

This will provide a usable input filter to the circuit. Measure your resulting inductances if possible. Calculate and wind the 56 nH inductor. This is much more critical, but a slightly wider capacitance range on the VCO tank circuit will allow proper tuning if you do not have exactly 56 nH.

    b. Ensure that your circuit is very "tight." No loose wires are allowable. Place several 0.1 μF capacitors between your power supply and ground (the 4.5 V is designed to be battery operated) to eliminate any feedback through the power supply.

    c. *Do not* try to measure the oscillation frequency of the VCO directly as you tune, since a frequency counter, oscilloscope, or spectrum analyzer (even with a 10:1 probe) will greatly reduce the actual oscillation frequency of the VCO and cause the circuit *not* to work. If you have a spectrum analyzer that operates in the FM range, you can see the tuning of the VCO by holding the spectrum analyzer lead a few inches from the VCO. This provides sufficient coupling to observe the VCO frequency as well as the frequency of the station that you are tuning to. Having a frequency counter, oscilloscope, or spectrum analyzer directly connected to the circuit anywhere will cause it not to work due to feedback of these high-frequency signals and change in the input impedance of that part of the circuit. This, of course, is one of the hazards of working at high frequencies!

    d. Note that your body capacitance can greatly affect the oscillation frequency of the VCO while you are tuning if you do not take care to isolate yourself from the circuit.

2. Tune the circuit slowly through its range and see what FM channels can be received.

Questions/Problems

1. Comment on your results and the difficulties you experienced working at these high frequencies.
2. What could you do to improve the efficiency of the tuning process?
3. Outline the engineering and manufacturing steps required to construct and manufacture a single-chip FM radio for commercial sale using this chip.

12

THE NE/SE 565
PHASE-LOCKED LOOP

INTRODUCTION

Phase-locked loops (PLLs) are used extensively in communications for modulation, demodulation, and generation of frequencies. IC PLLs can be used at relatively high frequencies, with general-purpose PLLs available for use up to about 10 MHz and more expensive, specialized PLLs available for higher frequencies. The purpose of this experiment is to introduce you to the terminology and operation of a common PLL. The NE/SE 565 can be used at up to 500 kHz and is a versatile, reliable, and predictable device. In this experiment you will adjust the frequency of the VCO built into the 565 PLL and determine the characteristics of the 565 PLL, which you will use in later experiments.

After completing this experiment, you will be able to

1. Build a 565 circuit to operate at any frequency with a desired lock range and capture range.
2. Understand how an FM input signal can be converted into an AM signal using the 565.
3. Understand the operation of a PLL.

REFERENCES

1. Young, Chapter 10.
2. Roddy and Coolen, Chapter 6.
3. Miller, Chapters 5 and 6.

4. Tomasi, Chapter 5.
5. Killen, Chapter 7.
6. Adamson, Chapter 13.

MATERIALS OR SPECIAL INSTRUMENTATION

Devices: NE/SE 565 PLL

Capacitors: 1 μF, 0.1 μF, 0.01 μF, two 0.001 μF

Resistors: 5 k potentiometer, two 0.68 k

THEORETICAL BACKGROUND

All PLLs, whether in discrete or integrated form, consist of three basic components: a phase detector, an amplifier, and a VCO. These are shown (internal to the 565) in Figure 12–1, which will be used in this experiment. Each of these three components has its own characteristics and is associated with a "gain" constant that describes its transfer characteristics. These constants are transfer functions that represent the output of the device divided by the input:

K_d = gain of the phase detector in volts/radian

K_a = gain of the amplifier (unitless)

K_0 = gain of the VCO in kilohertz/volt

The overall gain of the PLL is K_L and is the product of the three individual gains of the PLL components, as shown in Equation 12–1.

$$K_L = K_d K_a K_0 \qquad (12–1)$$

Initial Frequency Locking

The phase detector of a PLL compares the phase of the input signal with the phase of the VCO. The output is a voltage that is proportional to the phase difference between

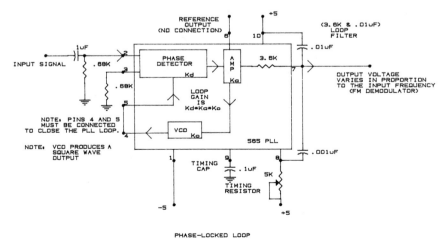

PHASE-LOCKED LOOP

FIGURE 12–1

the two signals; thus, K_d is in volts/radian. Figure 12–2 (upper left) shows a VCO (of a PLL) initially set at 1 kHz and two different input signals at lower (top) and higher (bottom) frequencies. If digital phase detection is used, an exclusive-OR (XOR) gate can serve as a phase detector and provides an easily understood example.

First, the lower frequency signal (at the upper left in Figure 12–2) is compared with the VCO signal, and the output of the XOR gate is a small dc value (after filtering). If, for example, the output of the phase detector is +1 V and the VCO input is biased at +2 V (with the loop open), this lower frequency input will reduce the free-running bias of the VCO and thus reduce the output frequency of the VCO. Similarly, if the higher frequency signal is compared with the VCO and the result is a larger dc value—+3 V, for example—this will tend to increase the frequency of the VCO. This type of arrangement will make the VCO frequency increase or decrease as required until it is the same as the frequency of the incoming signal. This frequency locking of the VCO to the incoming signal occurs very rapidly (generally on the order of $1/K_L$).

Maintaining Lock

To maintain the VCO at this new frequency, its input voltage must remain at the new level. After locking occurs, the VCO and the input are at the same frequency, so there is no frequency difference to compare. There is, however, now a phase difference between the input and the VCO due to the locking process. This phase difference is represented at the bottom of Figure 12–2 for the higher frequency signal after lock occurs. Notice that with the phase difference shown, the output of the XOR gate (which

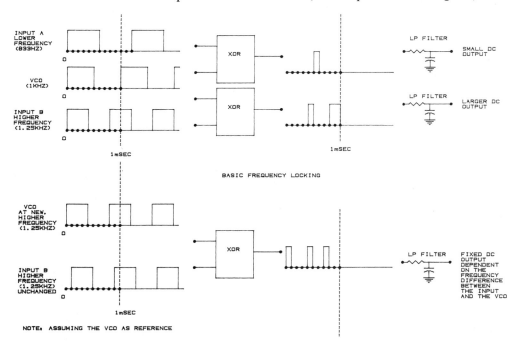

PHASE DETECTOR OPERATION

FIGURE 12–2

is now acting as a phase detector) has the same dc value (after filtering) as the initial XOR output at the top of the figure. Thus, the new VCO frequency is maintained by locking the phase of the signal so that the VCO frequency is locked to the frequency of the input signal.

General

The frequency and phase locking occur in a time of about $1/K_L$, so let's look at some typical values for the PLL. The value of $K_d K_a K_0$ for the 565 is given in Equation 12–2.*

$$K_d K_a K_0 = \frac{(33.6 f_0)}{V_c} \qquad \textbf{(12–2)}$$

where

f_0 = the free running frequency of the VCO

V_c = the total supply voltage

For example, for f_0 of 10 kHz and V_c of 10 V [as in Figure 12–1, $+5 - (-5) = 10$], $K_d K_a K_0 = 33,600$ Hz/rad, or 33.6 kHz/rad. Inverting this gives a locking time of about 29.8 μs.

Design formulas for the 565 are provided in the specifications in the Appendix for free-running frequency (of the VCO), lock range (PLL), and capture range (PLL). The capture range of the PLL is the range of frequencies onto which it will lock prior to being in lock. The lock range of the PLL is the range of frequencies in which the already-locked PLL will remain in lock. An example of use of the lock range formula is shown in Equation 12–3 using $f_0 = 10$ kHz and the supply voltage of 10 V again.

$$\text{Lock range:} \quad f_L = \pm \frac{8 f_0}{V_{CC}} = \frac{8 \times 1000}{10} = 8000 \qquad \textbf{(12–3)}$$

Using $C_2 = 0.01$ μF (as in Figure 12–1), the design formula in the specifications indicate that frequencies of 10 kHz ± 5.95 kHz can be captured, as shown in Equation 12–4. Of course, this is an approximate formula (as are all the design formulas provided), but it provides a good design tool if you know the modulation index of, for instance, an FM waveform.

$$\text{Capture range:} \quad f_c = \pm \left(\frac{1}{6.283} \right) \sqrt{\frac{6.283 \times f_L}{3.6 \times 10^3 \times C_2}} = 5.95 \text{ kHz} \qquad \textbf{(12–4)}$$

FM Demodulation Using a PLL

FM demodulation (actually, FM to AM conversion) can be accomplished using a PLL, as long as you can do the following:

1. Capture the center frequency of the FM signal. (Ensure this by adjusting the free-running frequency of the VCO to the center frequency.)
2. Remain in lock over the range of frequencies of the FM signal. If 1 and 2 are true, the amplitude of the voltage at the input to the VCO (pin 7 in Figure 12–1) will vary

---

* National Semiconductor LM565/LM565C applications information.

linearly with the input frequency changes. It is a straightforward process to convert an FM signal to an AM signal in this way; then simple AM demodulation methods can be used.

PROCEDURE AND QUESTIONS/PROBLEMS

Procedure

(Formulas are in the spec sheet in the Appendix.)

1. Calculate the free-running frequency range of the circuit shown in Figure 12–1. (HINT: Consider the range of the potentiometer.)
2. Calculate the lock range and capture range of the circuit for a 5 kHz free-running frequency. (Remember that V_{CC} is the total supply voltage of 10 V.)
3. Construct the circuit of Figure 12–1.
4. Open the loop by removing the short between pins 4 and 5. Measure the minimum and maximum free-running frequencies obtainable at the output of the VCO (pin 4) by varying the pot. Compare your results with your calculation from part 1.
5. Adjust the potentiometer to get a free-running frequency of 5 kHz at the output of pin 4.
6. Reconnect pins 4 and 5 to close the loop. Provide a 5 kHz squarewave from a signal generator (make this input frequency as close to the VCO free-running frequency as possible) at the circuit input (about 0.25 V peak).
7. Record the dc voltage at the output of the circuit with the PLL locked to the same frequency as the VCO free-running frequency (from 6).
8. Connect the oscilloscope to the output of the VCO (pin 4). Observe the oscilloscope while slowly increasing the frequency of the squarewave at the input. Record the frequency at which the PLL breaks lock (you will see a jittery waveform when it breaks lock instead of a clean squarewave).
9. Beginning at 5 kHz, slowly decrease the frequency of the input and determine the frequency at which the PLL breaks lock on the low end.
10. Compare your experimental lock range with your calculations.
11. With the oscilloscope still on pin 4 and the signal generator well above the capture range, attach the signal generator to the input. Reduce the frequency of the signal generator until you observe capture of the input signal by the PLL. Repeat from the low end of the frequency range.
12. Compare your experimental capture range with your calculations.
13. Measure the dc voltage at pin 7 for input frequencies of 4.5, 5.0, 5.5, and 6 kHz. Plot voltage (vertical) versus input frequency (horizontal). Sketch a straight line through the points. Calculate the reciprocal of the slope of this line in kilohertz/volt. Compare this with the theoretical value obtained using Equation 12–2.
14. Compare the phase difference between the input (using this as the 0° reference) and pin 7 for input frequencies of 4.5, 5.0, 5.5, and 6 kHz. Plot the phase difference (in radians) versus the voltage at pin 7. Sketch a straight line through these points and determine the slope in kilohertz/radian. This is the gain constant of the phase detector and amplifier combined.

83

Questions/Problems

1. Design a circuit with a free-running frequency of 200 kHz. Calculate the capture range and the lock range of this circuit.
2. Design (block diagram) a circuit to produce FM with a 400 kHz carrier using one 565 (HINT: use the VCO), an interface circuit to connect it to household ac wiring, and a receiver to demodulate the FM and receive a voice signal.

13

FSK MODULATION/DEMODULATION

INTRODUCTION

Frequency-shift keying (FSK) is the simplest form of FM because a carrier (usually a low-frequency carrier of several thousand hertz) is shifted between two discrete frequencies by a digital signal. FSK provided an early method of transmitting digital information over wires and phone lines, since one frequency could represent a 1 and the other frequency could represent a 0. It was used in many of the older 300-baud modems. FSK provides an interesting introduction to transmission of digital information. A good method of producing FSK would be to switch a VCO between the two desired output frequencies using an input digital signal with biasing. You will use a different method in this experiment to produce a simulated FSK signal and will use a 565 PLL for demodulation of the signal at the simulated receiver.

After completing this experiment, you will be able to

1. Understand FSK modulation and demodulation.
2. Construct an FSK modulator.
3. Construct a PLL, FSK demodulator.

REFERENCES

1. Young, Chapter 10.
2. Roddy and Coolen, Chapter 17.
3. Miller, Chapter 9.
4. Tomasi (*Advanced*), Chapter 1.
5. Killen, Chapter 13.
6. Adamson, Chapters 9 and 13.

MATERIALS OR SPECIAL INSTRUMENTATION

Devices: 555 timer, 565 PLL, 4052 analog switch, LM393 comparator, two LEDs, 556 dual timer (if desired to produce the 1100 and 1300 Hz frequencies for FSK)

Resistors: five 1 k, four 1.2 k, 4.7 k, two 10 k, two 47 k

Capacitors: (all in μF) 0.001, two 0.01, three 0.1, two 1, 22, 100

Potentiometers: 5 k, two 10 k, 20 k

THEORETICAL BACKGROUND

FSK modulation is usually used at relatively low frequencies. For instance, in the Bell (now AT&T) 103A asynchronous modem, 1070 and 1270 Hz signals are transmtited by the originator of the transmission, and 2025 and 2225 Hz signals are transmitted in reply by the receiver modem. This modem will operate up to 300 baud. Using two different sets of frequencies allows two-way transmission over a single two-wire line. The FM nature of an FSK signal is easily seen because of the variation in the frequency of the carrier by the signal. Because this is an FM signal, the modulation index must be considered in determining bandwidth. The formula for FM modulation index, m, is given in Equation 13–1.

$$m = \frac{\text{maximum deviation in frequency from center}}{\text{modulation frequency}} \qquad \textbf{(13–1)}$$

For an FSK signal that varies from 1100 to 1300 Hz (which are the frequencies used in the experiment), you can consider the center frequency to be 1200 Hz and the maximum deviation to be ± 100 Hz. The maximum modulation frequency of the 103A modem is 300 baud. Of course, this 300 baud refers to a digital signal (square-type wave) in place of the standard sine wave that is usually considered in FM modulation using this formula for modulation index. The fundamental frequency (using Fourier analysis) of a squarewave is the same as the frequency of the squarewave. The baud rate is the same as bits/second in this case, and 2 bits are contained in each period of a squarewave (a 1 and a 0). So we can consider the modulation frequency to be one-half the baud rate. With this factor, the modulation index becomes

$$m = \frac{100}{150} = 0.667$$

This is narrowband FM. From a Bessel function chart, the bandwidth of such a system is essentially twice the modulation frequency (of 150 Hz) or 300 Hz. In actuality, the bandwidth is more because of the harmonics of the squarewave signal.

Figure 13–1 shows the circuit to be used in this experiment. This looks like a complex circuit, but it is relatively straightforward, as you will see in the operational explanation that follows.

All the circuit components have been used in previous experiments except the LM393 comparator at the botton of the figure. The 555 at the upper left is used to produce a 1 to 25 Hz squarewave with a duty cycle of about 50 percent. This output simulates the digital signal that is switching the FSK output between its two frequencies. The 50 percent duty cycle is critical in an actual digital signal but is not critical for

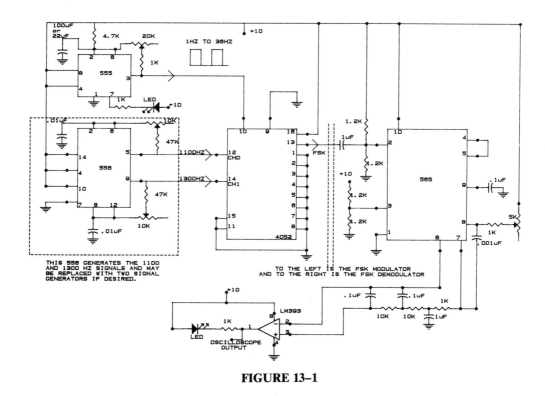

FIGURE 13–1

this experimental application. The LED will be used as one method of synchronizing the input (from the 555) and the demodulated output (of the 565). The 556 is optional (if you have two spare signal generators), since it just produces the 1100 and 1300 Hz signals representing the two FSK frequencies. You will use only two channels of the 4052 analog switch (since you need to switch between only two frequencies). The 1 to 25 Hz 555 timer output is used to switch the output of the 4052 analog switch between the two inputs (1100 and 1300 Hz) at channels 0 and 1. The result is an FSK signal that will look like a jittery sine wave on the oscilloscope but will not be recognizable in the time domain.

The two dashed lines between the 4052 and the 565 represent the wire or telephone lines over which the information is carried prior to demodulation. After arrival at the receiver, the FSK signal must be converted back into a digital signal. The 565 PLL is set to a free-running frequency of 1200 Hz, and the R and C values are set to ensure that both the 1100 and 1300 Hz frequencies are within the lock range (see formulas in Experiment 12 and the 565 specification sheet in the Appendix). The lock range for this experiment is ± 960 Hz, and the capture range is greater than ± 100. The 1.2 k resistors serve merely to allow the entire circuit to function from a single 10-V power supply. The output of pin 7 of the 565 is a voltage that switches between two levels at the rate of the input 555 timer. Because there are higher frequency components in this output signal and the voltage shift between 1100 and 1300 Hz input is very small, a method is required to reform the input (1 to 25 Hz) signal. The filter between pins 6 and 7 and the 393 is analyzed in Figure 13–2. As can be seen, this low-pass filter provides an attenua-

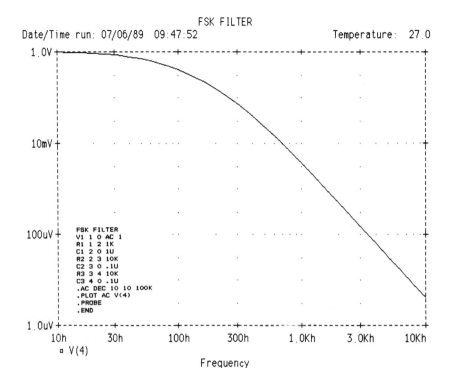

FIGURE 13–2

tion of over 200 to 1100 and 1300 Hz signals that may be in the output. The 393 comparator effectively amplifies the very small voltage difference between pins 6 and 7 to 0 and 10 V levels (nominal 10; actually, comparators provide a maximum output that is normally about 1.5 V less than the supply voltage). The output LED is used (along with the LED attached to the 555) as one method of synchronizing the output and input data.

PROCEDURE AND QUESTIONS/PROBLEMS

Procedure

(All references are to Figure 13–1.)

1. Construct the 555 circuit using the 100 μF capacitor, and adjust it to its minimum output frequency (about 1 Hz). Observe the output on the LED. If a brighter LED is needed, use less than 1 k with the LED.
2. Construct the 556 circuit (or set up two signal generators with squarewave outputs of 1100 and 1300 Hz, respectively).
3. Construct the 4052 circuit and attach the 555 and 556 (or signal generators) outputs to it. Observe the output waveform on an oscilloscope. At 1 Hz you should be able to observe the frequency shifting at the rate of flashing of the LED. Comment on the FSK signal.

88

4. Construct the 565 circuit. Adjust the free-running frequency of the 565 to approximately 1200 Hz (at pin 4) using the 5 k potentiometer with pins 4 and 5 not connected (open loop).
5. Finish construction of the 565/393 demodulation circuit as shown and connect the FSK input as shown.
6. While observing the two LEDs, adjust the 5 k potentiometer (pin 8 of the 565), if necessary, until the two LEDs are in sync. You can now observe FSK modulation and demodulation at a slow rate. (Of course, this is a squarewave, so you have just 1, 0, 1, 0, 1, 0, and so on.
7. Change the 555 timer capacitor to 22 μF, and adjust the 555 potentiometer to a frequency of 25 Hz. This corresponds to transmission of 50 baud. You will now have to sync the input and output using the oscilloscope, since the LEDs are too fast for your eyes. Sync the oscilloscope on the 555 output, and use a different channel to observe the output (pin 1 of the 393) or the demodulator at the same time. Sync the two signals by adjusting the 5 k potentiometer (on the 565) until the two signals are the same. (There may be a difference in amplitude and there may be a 180° phase shift.) The phase shift can be eliminated for syncing purposes by inverting one of the two channels of the oscilloscope.

Questions/Problems

1. Design (block diagram) a circuit to transmit and receive a 300-baud computer output signal from one part of your building to another.
2. Design an FSK modulator using the 565 to replace the 555 and 556 chips. (HINT: Use the VCO, not the entire PLL.)

14

DTMF DECODING

INTRODUCTION

Dual-tone multi-frequency (DTMF) signals are used in all Touch-Tone® telephones and are of widespread importance throughout the world. The purpose of this experiment is to introduce you to DTMF signaling, the use of PLLs to detect a DTMF signal, and a modern, single-chip DTMF receiver. In this experiment you will use two 567 PLLs to decode a single set of DTMF frequencies and then use a modern IC that is a complete DTMF receiver on a chip to detect the incoming dial signals on a Touch-Tone® line.

   After completing this experiment, you will be able to

1. Understand the operation of DTMF signal processing.
2. Construct a DTMF receiver using 567 PLLs.
3. Construct a DTMF receiver using a single IC.

REFERENCES

1. Young, Chapter 12.
2. Roddy and Coolen, Chapter 16.
3. Miller, Chapter 9.
4. Adamson, Chapter 13.

MATERIALS/SPECIAL INSTRUMENTATION

PLL Decoding

Two signal generators

Devices: two NE/SE 567 PLLs, LM741 op amp, NOR gate (TTL)

Resistors: five 1 k, two 20 k, 143 k (500 k potentiometer), 82 k (100 k potentiometer)

Capacitors: 0.1 μF, two 0.01 μF, two 10 μF, two 3.3 μF

DTMF Receiver

Devices: SSI 75T202 or RS 276-1303 DTMF receiver, colorburst crystal (3.579545 MHz), 74LS47 BCD to 7-segment decoder/driver, DL727 7-segment display, Touch-Tone® keypad (suggested but not required)

Resistors: 10 Ω, seven 330 Ω

THEORETICAL BACKGROUND

Tone Decoding with the 567 PLL

The basic operation of dual-tone multi-frequency (DTMF), or Touch-Tone®, telephone dialing is quite simple. Two specific frequencies are transmitted simultaneously each time a number is pressed. The chart shown in Figure 14–1 shows the standard tone signals and their corresponding alphanumeric meaning. For example, if a 5 is pressed, the frequency selected from the low-frequency group at the left of the chart is 770 Hz, and the frequency selected from the high-frequency group at the top of the chart is 1336 Hz. These two frequencies are transmitted simultaneously over the telephone line, and these are the two frequencies you will initially use in this experiment. The special function keys, * and #, are used for different functions on business and residential phones.

Figure 14–2 shows the use of the NE/SE567 PLL Tone Decoder to decode a single DTMF set of frequencies. This circuit is a minor modification of the circuit in the NE/SE567 specifications in the Appendix and includes a DTMF generator at the input. The DTMF generator consists of an op-amp summer network with two inputs, one at a low frequency selected from the left side of the chart in Figure 14–1 and one from the upper line of the chart. As you can see from the op-amp DTMF generator, it is easy to

	1209	1336	1477	1633
697	1	2/ABC	3/DEF	
770	4/GHI	5/JKL	6/MNO	SPECIAL FUNCTION KEYS
852	7/PRS	8/TUV	9/WXY	
941	*	0/OPER	#	

FREQUENCIES AND MEANING FOR TOUCH-TONE^R DIALING

FIGURE 14–1

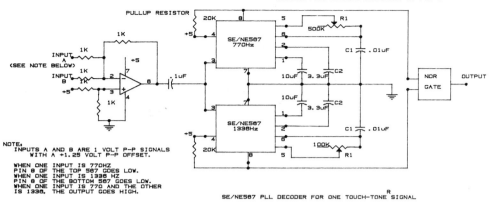

NOTE:
INPUTS A AND B ARE 1 VOLT P-P SIGNALS
WITH A +1.25 VOLT P-P OFFSET.

WHEN ONE INPUT IS 770HZ
PIN 8 OF THE TOP 567 GOES LOW.
WHEN ONE INPUT IS 1336 HZ
PIN 8 OF THE BOTTOM 567 GOES LOW.
WHEN ONE INPUT IS 770 AND THE OTHER
IS 1336, THE OUTPUT GOES HIGH.

SE/NE567 PLL DECODER FOR ONE TOUCH-TONE SIGNAL

FIGURE 14-2

generate a DTMF signal. Decoding of the signal is a matter of adjusting one of the 567 PLLs to lock onto one of the frequencies and the other to lock onto the other frequency.

Formulas to calculate the approximate free-running frequency of the VCO and the approximate bandwidth of the PLL (which corresponds to what was called the capture range of the PLL in the 565 experiments) are shown in Equations 14–1 and 14–2, respectively. These equations are taken directly from the specifications in the Appendix.

$$f_0 \text{ (free-running frequency)} = \frac{1}{1.1 R_1 C_1} \qquad (14\text{–}1)$$

where R_1 and C_1 are the timing elements, as shown in Figure 14–2.

$$\text{BW (loop capture range)} = 1070 \sqrt{\frac{V_1}{f_0 C_2}} \qquad (14\text{–}2)$$

where V_1 is the input rms voltage and C_2 (in μF) is shown in Figure 14–2.

From Equation 14–1 and the resistor and capacitor values for the upper 567 shown in Figure 14–2, the free-running frequency is 770 Hz. From Equation 14–2 (with $V_1 = 0.2$ mV rms), BW is approximately

$$\text{BW} = 1070 \sqrt{\frac{0.2}{770 \times 3.3}} = 9.5\%$$

$$9.5\% \times 770 = 73 \text{ Hz}$$

A better way of evaluating bandwidth (and the method used to find the value of C_2) is to determine the desired bandwidth and use this and the graph of the bandwidth versus input signal amplitude (see 567 specifications in the Appendix). From Figure 14–1 the maximum desired bandwidth for detection of the 770-Hz signal is halfway to the nearest frequency above and below 770 Hz. This is $(852 - 770)/2 = 41$ Hz on one side and $(770 - 697)/2 = 36$ Hz on the other. A bandwidth of 77 Hz (or less) is required in this case to ensure that the PLL doesn't lock onto the wrong tone. This is 10 percent

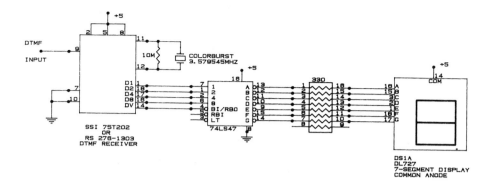

DUAL TONE MULTI-FREQUENCY RECEIVER WITH VISUAL OUTPUT

FIGURE 14-3

of f_0, so you follow the 10 percent line up to where it meets the curve (assuming a signal greater than 200 mV, where the curves become vertical). A value of 2.6×10^3 is shown. This is the value of $f_0 C_2$ required to provide this BW. Therefore, $C_2 = 2.6 \times 10^3 / 770 = 3.38 \ \mu F$ (3.3 μF is used).

A similar calculation for the 1336 Hz tone gives a required BW of 134 Hz, which, again, is a 10 percent BW, so the same 3.3 μF value can be used for C_2 in the second PLL.

DTMF Receiver

A block diagram of the DTMF receiver is shown in the Appendix. Referring to this block diagram, you can see that the demodulation is accomplished using band-pass filters (probably switched-capacitor filters) rather than PLLs in this single-chip design. Figure 14-3 shows the circuit to be constructed. This circuit consists of the receiver, a 7-segment-display decoder/driver that converts the binary output of the DTMF receiver into decimal equivalent codes, and the display device.

PROCEDURE AND QUESTIONS/PROBLEMS

Procedure

Tone Decoding With the 567 PLL

1. Construct the circuit of Figure 14-2. Note that an LED can be attached to the output of each 567 to determine lock. If this is done, the NOR gate is not required because if both are locked, then the NOR output will be high.
2. With no input and the frequency counter reading pin 5 of the upper 567, adjust the potentiometer (R_1) to get a free-running frequency of 770 Hz. Repeat with the lower 567 in the figure for a free-running frequency of 1336 Hz.
3. Connect a single-function generator to one of the inputs (1 V pk-pk sine wave with a +1.25 V offset) and a DMM (or oscilloscope) to pin 8 (the output of the upper 567). Slowly increase the frequency of the input from a low frequency up to a frequency at

which pin 8 goes low. Record this frequency, which marks the lower lock-in frequency of this PLL. Tune the function generator to a frequency above the lock-in range, and slowly decrease the frequency of the generator until the PLL locks in (pin 8 goes low). Record this frequency and determine the bandwidth of this PLL. Compare your measured bandwidth with the calculated bandwidth from the theoretical discussion. Comment on the difference.

4. Repeat steps 2 and 3 with the lower PLL for a frequency of 1336 Hz.
5. Test the operation of the decoder with both a 770 and a 1336 Hz input, and comment on the results.

DTMF Receiver

6. Construct the circuit of Figure 14–3.
7. Using the op-amp summer with 770 and 1336 Hz inputs, check the allowable bandwidth for the 770 Hz signal by varying the frequency of the 770 Hz signal to determine the upper and lower frequencies at which the receiver will indicate a 5 on the display. Comment on this bandwidth versus the bandwidth of the 567 circuit.
8. If a Touch-Tone® keypad is available, test the circuit using different inputs, and comment on the results.

Questions/Problems

1. Design a 567 decoder to detect the number 3.
2. Explain the operation of the complete Touch-Tone® decoder using the seven 567s shown in the specifications in the Appendix.
3. Explain (block diagram) how you could use a computer to record the numbers dialed from a Touch-Tone® keypad.

15

DIGITAL PLL FREQUENCY SYNTHESIZER

INTRODUCTION

Frequency synthesizers are used extensively in radio, television, and other applications where a specific frequency must be generated automatically. Digital frequency synthesizers provide a means for selecting a particular frequency using computer control or push-button input. Many communications devices have embedded microprocessors that can select a frequency or sweep through a frequency range to find a channel at the direction of the listener. Digital frequency synthesizers enable these devices to provide flexibility in commercial and military systems. The MC145106 is a CMOS phase-locked-loop frequency synthesizer for use over a wide range of frequencies. In this experiment you will construct a PLL based on the MC145106 and use it as a digital frequency synthesizer.

After completing this experiment, you will be able to

1. Understand the theory and operation of a digital PLL frequency synthesizer.
2. Construct a digital PLL frequency synthesizer for any desired frequency band.

REFERENCES

1. Young, Chapter 10.
2. Roddy and Coolen, Chapter 6.
3. Miller, Chapter 8.
4. Tomasi (*Fundamentals*), Chapter 5.
5. Killen, Chapter 7.
6. Adamson, Chapter 13.

MATERIALS OR SPECIAL INSTRUMENTATION

Devices: MC145106 (CMOS), 565 PLL, three 741 op amps, 8-pin spst DIP switch, colorburst crystal (3.57945-MHz), 2N3904-NPN (optional), LED (optional)

Resistors: 330 Ω, two 680 Ω, five 10 k, 68 k, 5 k potentiometer

Capacitors: two 20 pF, two 0.001 μF, 0.01 μF, two 0.047 μF

THEORETICAL BACKGROUND

The MC145106 is not a complete digital PLL, but it provides the divide-by-N digital input network, a crystal-controlled reference clock frequency, and a digital phase detector. The remainder of the PLL is external to the MC145106. Figure 15–1 shows the circuit to be used in this experiment. Since you are familiar with PLL operation, the remainder of this section will discuss the features and operation of the circuit.

MC145106 Operation

The internal operation of the MC145106 is depicted in block diagram form in Figure 15–1. The chip contains three fundamental parts:

1. The circuitry to divide the crystal frequency by 2 and then again by 512 or 1024. The crystal frequency is divided by a total of 1024 or 2048. The selection of divide-by-1024 or divide-by-2048 is made by opening (1024) or grounding (2048) pin 6.
2. The digital divide-by circuitry that takes the feedback signal at pin 2 and divides it by any number from 1 to 511 (there are nine input pins, and $2^9 = 512$).

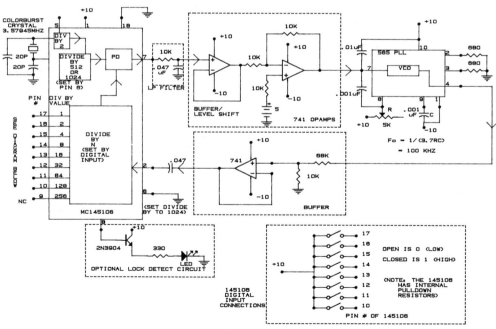

FIGURE 15–1

3. The digital phase detector that compares the crystal frequency (divided by 2048 in this experiment) with the incoming frequency at pin 2 (divided by N, set by the digital input). The output of the phase detector is a pulse-width-modulated (PWM) signal with a dc value proportional to the difference in frequency between the input signal (divided by N) and the crystal frequency (divided by 2048). The output of the phase detector is high when the input frequency (to the phase detector) is lower than the reference frequency (1747.8 Hz; see Equation 15–1) and low when the input frequency is higher than the reference frequency.

Circuit Operation

The MC145106 is used to control a VCO in a standard PLL configuration. The MC145106 acts as the phase detector, and the crystal-controlled input frequency acts as an input signal to the phase detector. The VCO output signal (after division by N in the MC145106) is compared with the crystal input frequency, which is calculated (for this experiment) in Equation 15–1.

$$\text{Phase detector crystal input} = \frac{3.57945 \times 10^6}{2048} = 1747.8 \text{ Hz} \qquad \textbf{(15–1)}$$

This is an important frequency that will be used later.

The 565 PLL is used as a convenient way of using an easily constructed VCO. Note that the PLL loop is open in the 565 (pins 4 and 5 are not connected), so *only the VCO portion of the 565 is being used*. For higher-frequency operation, the MC1648 VCO could be used (with slightly different voltage levels) in this circuit. Of course, the 741s would have to be replaced with higher-frequency devices.

The LP filter at the output of the MC145106 phase detector (pin 7) converts the PWM output signal into a 0 to 10 V dc signal (or close enough to dc to control the VCO).

The two 741 op amps between the LP filter and the 565 provide a buffer (unity-gain amp) and a level shifter to change the 0 to 10 V signal to a signal compatible with controlling the 565 VCO frequency.

The buffer between the 565 VCO and the input (pin 2) of the MC145106 changes the amplitude of the signal to the level (approximately 1 V pk-pk) that the MC145106 is capable of accepting. The 0.047 μF capacitor at the input to the MC145106 blocks dc.

Some Circuit Calculations

The VCO is initially adjusted to a free-running frequency of 100 kHz. To initiate the operation of the overall PLL, you will set the divide-by-N inputs to the MC145106 to a value that will make the input to the phase detector from the divide-by-N circuit the same as the input from the crystal. Equation 15–1 shows that the phase detector input from the crystal is 1747.8 Hz. (Yes, it is quite precise.) Thus, you need the other phase detector input also to be 1747.8 Hz. Equation 15–2 shows the equation to use for calculation of the VCO frequency or the divide-by-N value.

$$\frac{\text{Frequency of VCO (pin 4)}}{\text{Divide-by-}N \text{ value}} = 1747.8 \qquad \textbf{(15–2)}$$

For an input frequency of 100 kHz the required divide-by-N value is $(100 \times 10^3)/ 1747.8 = 57.21$. Since the divide-by-$N$ circuit is in increments of 1, you will set it at 58 initially. This will provide a calculated frequency of 1747.8 × 58 = 101.37 kHz. Previous laboratory experiments using these values gave an output of 101.38 kHz on the frequency counter, so these calculations are quite accurate.

The limits on the bandwidth of this VCO under these conditions are about 40 to 275 kHz, which, using Equation 15–2, provides a limit on usable divide-by-N values of about 23 to 157. If a divide-by-N value of 28 is used, a frequency of 48.94 kHz is synthesized, and if a divide-by-N value of 128 is used, a frequency of 223.73 is produced.

PROCEDURE AND QUESTIONS/PROBLEMS

Procedure

(NOTE: Do not connect the entire PLL circuit until step 8.)

MC145106

1. Construct the MC145106 portion of the circuit. To test operation:
   a. Check that the output of pin 5 is exactly one-half of the colorburst crystal frequency using a frequency counter.
   b. Provide a 100 kHz, 1 V pk-pk squarewave signal to pin 2 of the MC145106 through the 0.047 μF capacitor. Check the dc output of pin 7 of the MC145106. Note that the output of pin 7 is PWM, and the average output varies depending on the difference between the input frequency and the preset, crystal-controlled frequency. Observe the output of pin 7 on the oscilloscope, but this will not provide useful information. The key is the dc value of this waveform, which can be measured using a DMM. Record this dc value.
   c. Show how you calculate the frequencies into the phase detector.
   d. Change the frequency into pin 2 to 50 kHz and then 200 kHz and record the dc outputs at pin 7. Is this output linear when plotted versus input frequency? Comment on the results.
   e. Discuss the operation of this device. Note that pin 7 goes high when the frequency out of the divide-by-N circuit is less than the constant frequency (1747.8 Hz) generated by the crystal circuit.

565 VCO

2. Construct the 565 portion of the circuit. Set the free-running frequency of the VCO to 100 kHz by adjusting the 5 k potentiometer.
   a. Test the operation of this portion of the PLL by varying the input voltage of pin 7 of the 565 and observing the resulting change in output frequency of the VCO. Record the input voltages required to vary the frequency to 50 kHz and to 200 kHz. Comment on the results.
   b. Discuss the operation of this portion of the PLL.

Buffers and LP Filter

3. Construct the lower, single-op-amp buffer portion of the circuit (between the 565 output and the MC145106 input). Do not connect it to either the 145106 or the 565 until you are sure it is operating properly. It is simply a voltage divider and a buffer.
   a. Test the operation of this circuit by connecting the 565 to the buffer input and ensuring that the output of the buffer is about 1 to 1.5 V pk-pk.

b. Connect the buffer output to the input of the 145106, and change the divide-by-N value while observing the dc value at pin 7 of the 145106. This dc value should increase for divide-by-N values above 58 and decrease for values below 58. Show that you understand these results.

4. Construct and connect the LP filter. Calculate the cutoff frequency of this filter and discuss why it was set at this value.

5. Construct the upper, two-op-amp buffer/level shifter. Do not connect either end of this circuit into the overall PLL until the buffer/level shifter is fully operational.

   a. This circuit should cause a 0 V input to become a high output and a $+10$ V input to become a low output, which is inversion and level shifting.

   b. Discuss the operation of this circuit.

6. Connect the entire PLL circuit *except* pin 7 of the 565.

   a. Ensure that the entire circuit is working without connecting to pin 7 of the 565 by changing the 565 free-running frequency and noting the dc voltage value that occurs at the input to the 565.

   b. Discuss the result of this test and your understanding of the operation of the circuit.

Completing the Circuit

7. Reset the VCO free-running frequency to 100 kHz. Set the divide-by-N value to 58. Explain why this number is significant.

   (NOTE: You may need to break the circuit, which resets it, if the value of N that you set is too high or low. Keep the connection to pin 7 of the 565 easily accessible, since this is a good place to break the loop.)

8. Connect the complete PLL circuit.

   a. Vary the divide-by-N input to 10 different values within the operating range of the 565 (approximately 23 to 157), and make a chart of the input digital number, the calculated frequency, and the actual output frequency from a frequency counter. Make a graph of this chart.

   b. Comment on the linearity of the resulting graph.

   c. Determine the limits of the frequency ranges available by changing the divide-by-N number until the calculations no longer match the actual output of the VCO. Record and comment on these limits.

Questions/Problems

1. Design a similar digital frequency synthesizer using higher output frequencies with the MC1648 VCO.

2. Calculate the divide-by-N value required to produce a 1.190 MHz frequency using a 10.24 MHz crystal on the 145106 with a divide-by-1024 crystal circuit. What is the range of frequencies directly usable with this configuration?

16

PULSE-AMPLITUDE MODULATION AND SAMPLE AND HOLD

INTRODUCTION

Pulse-amplitude modulation (PAM) is the basic method of sampling analog quantities prior to using an analog-to-digital converter (ADC). A sample-and-hold (S&H) device uses PAM to sample an analog signal and then holds the voltage level of the sample steady during the time required for the ADC to change the input level into a digital representation. Many more-expensive ADCs contain a S&H on the chip. Thus, a discussion of PAM leads directly into S&H, and as this experiment shows, a basic method of performing the PAM function leads directly into a S&H circuit.

After completing this experiment, you will be able to

1. Understand the principles of pulse amplitude modulation and sample-and-hold circuits.
2. Construct a basic PAM circuit.
3. Convert the PAM circuit to an S&H circuit.

REFERENCES

1. Young, Chapter 11.
2. Roddy and Coolen, Chapter 11.
3. Miller, Chapter 9.
4. Tomasi (*Advanced*), Chapter 4.
5. Killen, Chapters 11 and 12.
6. Adamson, Chapter 8.

MATERIALS OR SPECIAL INSTRUMENTATION

Devices: LM555 timer, LM1458 dual op amp (or two LM741 op amps), MPF102 *n*-channel JFET

Resistors: 1 k, 2.7 k, 10 k, 1 M

Capacitors: 0.01 μF, 0.1 μF

THEORETICAL BACKGROUND

The Nyquist sampling theorem states that in order to recover the original signal samples of that signal, the signal must be sampled at a minimum of twice the maximum frequency of the signal. For example, a 200 Hz signal must be sampled at a minimum of 400 Hz (400 samples per second). In actual work, even the telephone company (using state-of-the-art equipment) samples 3400 Hz (bandwidth) voice signals at 8000 Hz, which is about 2.35 times the maximum signal frequency. In this experiment you will initially sample at about five times the signal frequency and observe the effect of changing the sample rate.

PAM is sampling of a signal so that the output of the signal follows a small portion of the original signal. Figure 16–1 shows a 200 Hz sine wave with a dc offset (like the input to your circuit) and the sampled output produced by a 1 kHz PAM circuit with a 20 percent duty cycle. Note that if you superimpose the input sine wave onto the sampled output signal from your circuit, this is the waveform you should see. If you

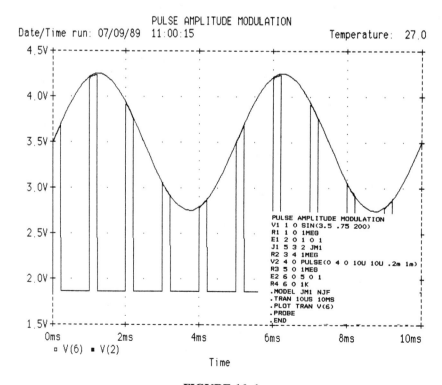

FIGURE 16–1

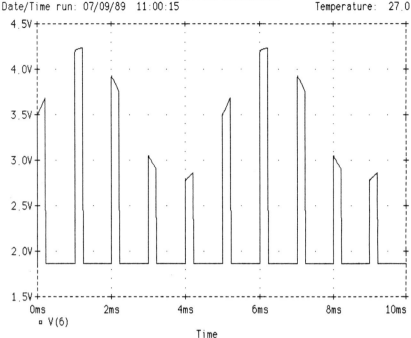

□ V(6)

Time

FIGURE 16–2

look at the sampled output shown in Figure 16–2, you can see that pieces of the original sinusoidal signal have been ''sampled'' and, perhaps, understand that the original signal can be recovered from the samples. Reconstruction of the original sinusoid from the samples of Figure 16–2 can be done using an integrator circuit that would ''fill in the gaps'' between the samples. Another method of reconstruction is to use a low-pass filter that passes only frequencies of 200 Hz (the original frequency) and below. The frequency spectrum of the sampled signal of Figure 16–2 is shown in Figure 16–3. If you compare the time waveform with the frequency spectrum, the following important points emerge:

1. There is a large dc component in the frequency spectrum (the dc component is easily seen in the time waveform).
2. There is a component of the sampled signal at 200 Hz, which is the original signal frequency.
3. The higher frequency components are due to the sampled pulses at 1000 Hz. These higher frequency components look like a carrier with 200 Hz sidebands at 1 kHz and all harmonics. Thus, PAM represents amplitude modulation using a pulse rather than a single carrier frequency.
4. A sharp low-pass filter will eliminate all the higher frequency signals, and only the original 200 Hz (+dc) will remain.

A S&H circuit takes samples of the signal exactly like the waveform of Figure 16–2 but includes a circuit (usually a capacitor) to ensure that the output of the circuit

105

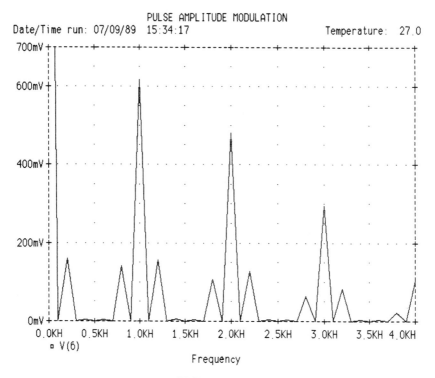

700mV

600mV

400mV

200mV

0mV

0.0KH   0.5KH   1.0KH   1.5KH   2.0KH   2.5KH   3.0KH   3.5KH 4.0KH

□ V(6)

Frequency

FIGURE 16–3

remains at the peak value of the sample until the next sample occurs. If you compare Figure 16–2 with Figure 16–4, you will see that the sampling still occurs and the output follows the input during the sampling time (20 percent of the period of the pulse), but the voltage is then held at this level until the next sample occurs. This is important to A/D conversion so that the A/D device has a steady value during its conversion time. You will see the need for this in Experiment 17 on A/D conversion.

Figure 16–5 contains the circuit for this experiment. For the PAM portion, you will use the entire circuit except the 0.01 μF capacitor. This capacitor will be all that is added to convert the circuit to a S&H. The 555 timer is used to produce an approximately 20 percent duty cycle pulse that acts as the sampling pulse. An n-channel JFET is used as a simple switch, so that the source-to-drain connection is made when the 555 pulse is high (about 4.5 V) and is broken when the pulse is low (0 V). The two op amps are used to buffer the input and output. Note that a 1458 or two 741s can be used with good results at these frequencies, but a JFET input (higher input impedance) op amp like the TL082 will provide a better buffer and cause less droop in the S&H waveform. *Droop* is the slowly falling exponential seen in Figure 16–4 at the top of the holding waveform. With a 741 (about 1 MΩ input resistance), there will be a greater falloff (droop) than with a TL082 (about 10 MΩ input resistance).

For PAM (without the 0.1 μF capacitor), circuit operation is straightforward. The input is a 1.5 V pk-pk sinusoid with a +3.5 V dc offset. The offset is required to bias the JFET off except when a positive pulse arrives from the 555. Other methods of biasing

106

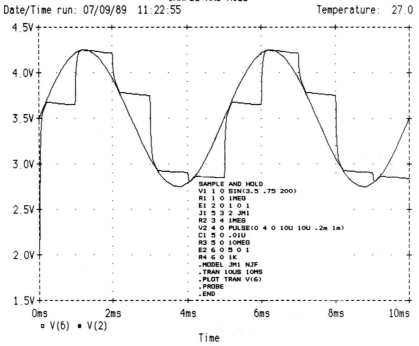

SAMPLE AND HOLD

Date/Time run: 07/09/89 11:22:55   Temperature: 27.0

```
SAMPLE AND HOLD
V1 1 0 SIN(3.5 .75 200)
R1 1 0 1MEG
E1 2 0 1 0 1
J1 5 3 2 JM1
R2 3 4 1MEG
V2 4 0 PULSE(0 4 0 10U 10U .2m 1m)
C1 5 0 .01U
R3 5 0 10MEG
E2 6 0 5 0 1
R4 6 0 1K
.MODEL JM1 NJF
.TRAN 10US 10MS
.PLOT TRAN V(6)
.PROBE
.END
```

□ V(6) ■ V(2)

Time

FIGURE 16–4

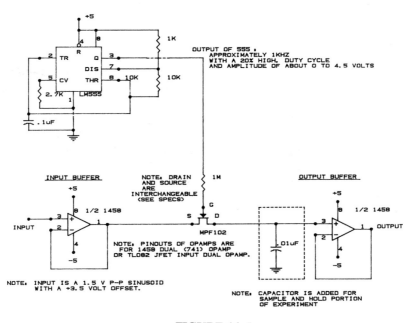

FIGURE 16–5

the JFET are certainly possible, but this is perhaps the simplest. The JFET turns on for 20 percent of its 1 ms period, and during this 0.2 ms the input waveform is allowed through to the output. A waveform like that of Figure 16–2 is the result.

For a S&H (with the 0.1 μF capacitor) the same switching occurs, except during the 0.2 ms that the input waveform is connected to the capacitor (and the output), the capacitor charges to the voltage of the input waveform (a fast charging time is required). When the input is disconnected from the capacitor, it stores the charge, and the output voltage remains at the sample level except for droop due to exponential discharge of the capacitor into the input impedance of the output op amp.

PROCEDURE AND QUESTIONS/PROBLEMS

Procedure

PAM

1. Construct the 555 circuit. Measure and record the frequency and the duty cycle of the output.
2. Construct the circuit of Figure 16–5 without the 0.01 μF capacitor.
3. Using the input shown (1.5-V pk-pk sine with a +3.5 V offset), observe and sketch the output of the circuit for a 200 Hz input sinusoid.
4. Increase the frequency of the input sinusoid (the amplitude and offset remain the same) to 400 Hz, 500 Hz (the Nyquist frequency), and 600 Hz. Comment on the sampling waveforms, your understanding of the Nyquist criteria, and the ability to generate the original waveform from the samples at each of these frequencies.

S&H

5. Place the 0.01 μF capacitor in the circuit, as shown in Figure 16–5.
6. Repeat step 3.

Questions/Problems

1. For the S&H circuit, if the output impedance of the first op amp is 1 Ω and the drain-source resistance of the JFET is 99 Ω, what is the time constant for charging the 0.01 μF capacitor?
2. For the S&H circuit, if the input impedance of the second op amp is 1 MΩ and the JFET has no leakage, what is the discharge time constant of the holding circuit?
3. The circuit used in Experiment 9 on time-division multiplexing also sampled a signal (several signals). Discuss the differences and similarities between the circuits.

17

A SUCCESSIVE-APPROXIMATION ANALOG-TO-DIGITAL CONVERTER

INTRODUCTION

Analog-to-digital converters (ADCs) are used in all computer applications that require interface with the analog world. In addition, if a digital readout of an analog measurement is needed, a stand-alone ADC can be used to generate the bit patterns required to drive a decimal or hexadecimal display. The method of successive approximation is widely used because of its accuracy (generally $\pm$ 1/2 LSB) and speed (the number of decisions to be made equals the number of output bits of the ADC). In this experiment the ADC0808 or ADC0809 is used in a stand-alone ADC configuration.

After completing this experiment you will be able to

1. Understand the operation of a successive-approximation ADC.
2. Understand the operation of the ADC0808/0809.
3. Build an ADC0808/0809 stand-alone circuit.
4. Use the ADC0808/0809 in other A/D applications.

REFERENCES

1. Young, Chapter 12.
2. Roddy and Coolen, Chapter 17.
3. Miller, Chapter 9.
4. Tomasi (*Advanced*), Chapters 1 and 4.
5. Killen, Chapter 12.
6. Adamson, Chapter 9.

MATERIALS OR SPECIAL INSTRUMENTATION

Devices: ADC0808/0809 (The 0808 has ±1/2 LSB resolution and the 0809 has ±1 LSB resolution), 500 kHz to 1 MHz TTL oscillator (not required; you may use a signal generator), eight LEDs

Resistor: 5 k potentiometer

Capacitor: 0.1 μF

THEORETICAL BACKGROUND

General

Several important items provided in specifications of ADCs are resolution, error, clock frequency, and conversion time. For the ADC0808/0809 (see specifications in the Appendix), the resolution is 8 bits; the error is ±1/2 LSB (0808) and ±1 LSB (0809); the allowed clock frequencies are 10 kHz to 1280 kHz; and the conversion time is 100 μs(typical) or 116 μs(max), with a 640-kHz clock.

An 8-bit resolution tells you that $2^8 = 256$ levels are available and that the LSB is $V_{max}/256$ V. In this case, using a 5 V supply, with a 0 to 5 V input range, $5/256 = 0.0195$ V. So each of the 256 bits represents 0.01953 V, with 0000 0000 (00H) representing 0.000 V and 1111 1111 (FFH) representing $(255/256) \times 5 = 4.9805$ V.

The specified error of ±1/2 for the 0808 (or 1 for the 0809) means the maximum error in the output is $\pm(0.0195/2) = 0.00977$ V for the 0808 (or 0.0195 V for the 0809).

The conversion time tells you how long it takes the device to complete conversion of the analog input to the digital output. During this period of time (about 100 μs here), the input must remain steady. This is why sample-and-hold (S&H) devices (such as the one constructed in Experiment 16) are used prior to ADCs. Sample-and-hold amplifier ICs are available and many more-expensive ADCs have S&H circuits built into the ADC. The conversion time of 100 μs can be used for a free-running ADC like the one you will construct in this experiment; however, if you are enabling the output of the ADC as part of a computer program, you must ensure that the ADC is ready for the transfer and must not request input faster than the maximum conversion time specified, which is 116 μs. This is usually accomplished by having the ADC generate an interrupt to the computer when it has completed the conversion and is ready for the computer to read the output bits. The conversion times are specified at 640 kHz clock frequency, but do not assume that a 1280 kHz clock frequency will make the conversion times twice as fast (unless specified by the manufacturer). The various allowable clock frequencies enable you to use readily available clock pulses as long as they are in the specified range of 10 kHz to 1280 kHz. Slower clock speeds *will* slow down the conversion process, since each bit converted is related to the clock speed.

The decision process used in a successive-approximation ADC is illustrated in Figure 17-1 for a 0 to 5 V input, 8-bit ADC (similar to the ADC0808/0809). In the figure the digital approximation for an analog input of 2.120 V is developed. Details of the conversion process are contained in Figure 17–1, with some additional comments provided.

1. The first step is to compare the input with one-half the supply voltage (represented by (1000 0000), or 80H = 2.500 V). If the input is higher, the first (MSB) bit is set to 1. If the input is lower, the MSB is set to 0, as is done here.

110

INPUT: 2.120 VOLTS

DIGITAL OUTPUT:
0110 1101
MSB   LSB

TRY #	BINARY	HEX	VOLTS		OUTPUT BITS	PROCESS
1.	10000000	(80H)	- 2.5	TO HIGH	0	FIRST BIT IS 0 TRY NEXT BIT = 1
2.	01000000	(40H)	- 1.25	TO LOW	1	SECOND BIT IS 1 TRY NEXT BIT = 1
3.	01100000	(60H)	- 1.875	TO LOW	1	THIRD BIT IS 1 TRY NEXT BIT = 1
4.	01110000	(70H)	- 2.188	TO HIGH	0	FOURTH BIT IS 0 TRY NEXT BIT = 1
5.	01101000	(68H)	- 2.031	TO LOW	1	FIFTH BIT IS 1 TRY NEXT BIT = 1
6.	01101100	(6CH)	- 2.109	TO LOW	1	SIXTH BIT IS 1 TRY NEXT BIT = 1
7.	01101110	(6EH)	- 2.148	TO HIGH	0	SEVENTH BIT IS 0 TRY NEXT BIT = 1
8.	01101101	(6DH)	- 2.129	--------	1	A DECISION LEVEL MUST EXIST HERE TO ENABLE A + OR - 1/2 LSB MAX ERROR

8-BIT SUCCESSIVE APPROXIMATION EXAMPLE

FIGURE 17-1

2. The next MSB is now set to 1 (0100 0000) = 1.25 V, and this voltage is compared with the input. Since this is lower than the input, this bit remains set to 1, and the next bit is set and compared as shown in the figure.

3. This process is continued until all 8 bits have been determined, and the final output (01101101 = (6DH) = 2.129 V) is then enabled, as shown in the figure.

4. The error for this example is approximately 0.009 V, which is within the specified ±1/2 LSB.

During this successive approximation process, the output of the ADC is not available to the output device (computer or display), so if a computer program calls for an input during the conversion process, a read error (or a more difficult to specify error) will occur in the program, or an incorrect value will be read.

The Circuit (Figure 17-2)

The 0.1 µF capacitor is just for decoupling to eliminate possible feedback oscillation. The 5 k potentiometer provides a range of input values for you to observe the digital output and verify that it matches the measured analog input voltage. The ADC0808/

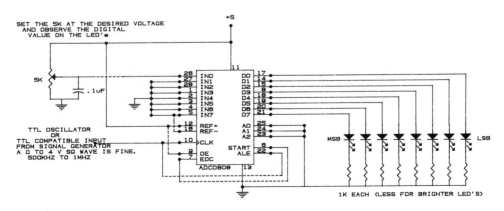

FIGURE 17-2

0809 has many capabilities not used in this experiment; a discussion of the wiring of the circuit will highlight some of the more sophisticated aspects of the device.

The pins labeled IN_0 to IN_7 allow eight separate analog inputs to the device. These are time-division multiplexed by the pins labeled A_0, A_1, and A_2 in Figure 17–2. A_0 is the LSB, and A_2 is the MSB, so that 000 activates input 0 (IN_0), 001 activates IN_1, 101 activates IN_5, and so forth. We simply wire the three control pins to ground, which activates only one input (IN_0) for this experiment.

The REF+ and REF− pins in Figure 17–2 provide the upper and lower voltage range to the 256-resistor ladder that is used for the comparisons. For accuracy, these reference pins must be stable and may not be tied directly to the supply voltage, since it may tend to drift. As long as you wish your comparisons to match your supply voltage, the REF+ can be tied to V_{CC}, and REF− can be grounded as is done here. This does reduce the accuracy if the supply voltage drifts, but you will find that the output bits represent the input voltage quite accurately in this experiment.

The OE (output enable), EOC (end of conversion), START, and ALE (address latch enable) pins are generally used to control the timing of the information transfer using a microcomputer. The EOC output is used to interrupt the computer program when a conversion has been completed. When the computer is ready to read the output, it enables the OE pin, reads the 8 bits, and then enables the ALE and START pins to begin the process again. Of course, if multiplexing of more than one input is desired, the A_0, A_1, and A_2 pins must be set each time prior to enabling the ALE and START pins.

For this experiment

1. The OE pin is tied high (to the supply), which allows continuous operation (as fast as the conversion time allows).
2. The ALE is tied to the clock, which enables the input each clock cycle. This means that as soon as a conversion is completed, on the next clock cycle, the input is enabled.
3. The EOC and START are tied together. This means that as soon as the conversion is completed, the START is enabled and a new cycle is begun.
4. These three connections allow continuous operation of the ADC0808/0809.

PROCEDURE AND QUESTIONS/PROBLEMS

Procedure

Save this circuit for use in Experiment 18.

1. Construct the circuit of Figure 17–2. Use LEDs requiring little drive current, such as an 8-LED package, or else you may need a buffer between the output and the LEDs.
2. Set the input at the voltages shown in Table 17–1, and fill in the remainder of the table. Show how you arrived at the calculated output values. The calculated output can be found as in Figure 17–1 or you may calculate the nearest 8-bit number to the analog voltage (using normal arithmetic rounding).

TABLE 17–1

Voltage	Calculated Output	Actual Output
0.00 (GND)	0000 0000 (00H)	
1.00		
2.12	0110 1101 (6DH)	
2.50	1000 0000 (80H)	
3.50		
4.50		
5.00 (V_{CC})	1111 1111 (FFH)	

3. Show how you would change the wiring of Figure 17–2 to read any of the eight inputs using three pins of a DIP switch at pins A_0, A_1, and A_2.

Questions/Problems

1. Design a block diagram of a circuit to use an S&H amplifier prior to each of the eight inputs to the ADC and to control the ADC using a microprocessor. List the timing factors involved in this process.

18

DIGITAL-TO-ANALOG CONVERSION

INTRODUCTION

Digital to analog (D/A) conversion is necessary in all computer applications to enable the computer to communicate with or control real-world analog devices. The purpose of this experiment is to introduce you to a common 8-bit digital-to-analog converter (DAC), the DAC0808. You will observe its accuracy, and then connect it to the ADC0808 constructed in Experiment 17 and observe the overall performance of both devices.

After completing this experiment, you will be able to

1. Understand the operation of an R-$2R$ ladder, current-source DAC.
2. Build a DAC using the DAC0808.
3. Use the DAC0808 with the ADC0808.
4. Use the DAC0808 in other D/A applications.

REFERENCES

1. Young, Chapter 12.
2. Roddy and Coolen, Chapter 17.
3. Miller, Chapter 9.
4. Tomasi (*Advanced*), Chapters 1 and 4.
5. Killen, Chapter 12.
6. Adamson, Chapter 9.

MATERIALS OR SPECIAL INSTRUMENTATION

Devices: DAC0808, 741 op amp

Resistors: two 10 k

Capacitor: 0.1 μF

THEORETICAL BACKGROUND

A block diagram of the DAC0808 is shown in the Appendix. Note that in the block diagram are current switches, an R-$2R$ ladder network, a bias circuit, and a pair of transistors that act as a current source. The heart of the DAC is the R-$2R$ ladder.

For an explanation of the operation of an R-$2R$ ladder, refer to Figure 18–1, which shows a 4-bit R-$2R$ ladder network for D/A conversion. The switches are connected either to ground (if that particular input bit is a 0) or to the op-amp input, a virtual ground, (if that input bit is a 1). For impedance purposes, the base of each $2R$ resistor is always connected to ground. Thus, the impedance seen to the right of each voltage junction (at the top of the network) is R. From basic voltage division, the voltages at junctions to the right of the reference voltage, $-V$, are each one-half of the previous voltage. This is the same for the 8-bit DAC0808 as for this 4-bit example circuit. For the example of Figure 18–1, the digital input is 1001. This closes the switches on the MSB and the LSB and opens the other two switches. The same current flows at the reference for any switch position; however, the current to the output is a function of which switches are connected to the output and, hence, a function of the digital input of the device. In Figure 18–1 the total current at the reference voltage is always V/R, but the current to the output is the sum of only the two currents due to the two digital 1 inputs. Therefore, the current from the output is $V/2R + V/16R = (9/16)(V/R)$, which corresponds directly to the 1001 digital input. The op amp with a feedback resistor R_f is a current-to-voltage converter. If $R_f = R$ (from the resistor ladder), the analog output voltage is $R(9/16)(V/R) = (9/16)V$. If the desired output voltage range of the DAC is 5 V and the reference voltage is 5 V, the output of this 4-bit converter would be $(9/16)5$ V = 2.81 V, which corresponds to the digital value of 1001.

The actual circuit used for D/A conversion is shown in Figure 18–2; it corresponds very closely to the theoretical circuit discussed. If the values of the 10.0 k resistor attached to pin 14 and the 10.0 k feedback resistor on the op amp are *exactly* the same, the D/A conversion will be quite precise.

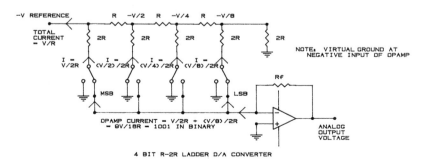

4 BIT R-2R LADDER D/A CONVERTER

FIGURE 18–1

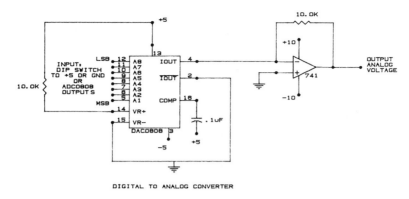

DIGITAL TO ANALOG CONVERTER

FIGURE 18–2

PROCEDURE AND QUESTIONS/PROBLEMS

Procedure

1. Construct the circuit of Figure 18–2.

2. Using 8-DIP switch inputs with the switches connected either to ground (closed) or +5 V (open), set the input to the values in Table 18–1 and complete the table. Comment on the accuracy of the DAC.

TABLE 18–1

| Digital Input | Analog Output | |
	Theoretical	Measured
0000 0000	0.000 V	
1000 0000	2.500 V	
1111 1111	4.980 V	
0100 0000		
1100 0000		
1100 0001		
0000 0001		

TABLE 18–2

Analog Input (V)	ADC0808 Output	DAC0808 Output
0.000		
1.000		
2.500		
3.000		
4.000		
5.000		

3. Remove the 8-DIP switch input and connect the output of the ADC0808 constructed in Experiment 17 to the inputs of the DAC0808. You do not have to remove the LEDs on the ADC.
4. Set the input to the ADC to the analog voltage values shown in Table 18–2. Complete Table 18–2 and comment on the accuracy of the combination ADC and DAC.

Questions/Problems

1. Sketch and analyze the output of an 8-bit R-$2R$ ladder network similar to Figure 18–1 with an input of 1100 1000.
2. Determine the output of a 5 V, 10-bit, R-$2R$, DAC with an input of 10 1000 1000. What is the accuracy of this device, in volts, (assuming that the computer has 10-bit accuracy)?
3. Determine the output, in volts, of a 5 V, 16-bit, R-$2R$ ladder network DAC with an input of AF08 (HEX).

19

PULSE-WIDTH AND PULSE-POSITION MODULATION

INTRODUCTION

Pulse-width modulation (PWM) (also called pulse-duration modulation, or PDM) is used in communications applications and is widely used in motor control. Pulse-position modulation is also a useful, but not widely used, communication technique. The purpose of this experiment is to introduce you to a method of producing and observing PWM and PPM waveforms and methods of demodulation.

After completing this experiment, you will be able to

1. Understand PWM and PPM.
2. Build a 555 circuit or a comparator circuit to produce PWM.
3. Build a 555 circuit to produce PPM.
4. Design a circuit to convert from PWM to PPM.
5. Design a circuit to convert from PPM to PWM.
6. Design a circuit to demodulate PWM.

REFERENCES

1. Young, Chapter 11.
2. Roddy and Coolen, Chapter 11.
3. Miller, Chapter 9.
4. Tomasi (*Advanced*), Chapter 4.
5. Killen, Chapter 11.
6. Adamson, Chapter 8.

MATERIALS OR SPECIAL INSTRUMENTATION

Two function generators (or one function generator and a clock input, which can be made from another LM555 similar to the 555 circuit in Experiment 13)

Devices: LM555 timer

Resistors: 3 k, 3.9 k, 9.1 k

Capacitor: 0.01 μF

THEORETICAL BACKGROUND

PWM

A basic method of producing PWM (and PPM) is shown in block diagram form in Figure 19–1. Figure 19–2 shows the waveforms at the input and the output of the comparator. In Figure 19–2 the analog input to the positive side of the comparator is the sinusoid, and the sawtooth waveform (which acts as a clock) is the input to the negative side of the comparator. Note, in Figure 19–2, that the sawtooth clock has a period of 0.2 s and the sine wave, a period of 1 s. Nyquist's criteria from Experiment 16 apply, since this is just another form of sampling. In Figure 19–2 the sampling rate is five times the frequency of the analog waveform. The PWM output shown in Figure 19–2 occurs because the output of the comparator goes high when the sawtooth clock first starts. This occurs because the analog input voltage is *always* higher than the sawtooth at the beginning of the period of the sawtooth. The output, PWM, pulse goes low when the sawtooth voltage is greater than the analog input.

Figure 19–3 shows the frequency spectrum of the PWM waveform in Figure 19–2. The sampling is at 5 Hz, and the analog input is 1 Hz, but Figure 19–3 shows that

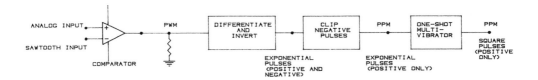

```
PWM
V1 1 0 SIN(100 100 1)
V2 1 2 PWL(0 0 .199 200 .2 0 .399 200 .4 0 .599 200 .6 0 .799 200 .8 0 .999 200
1 0)
R1 2 0 10K
V3 3 0 DC 100
S1 3 4 2 0 RELAY
R2 4 0 1K
.MODEL RELAY VSWITCH
.TRAN 10mS 1S
.PLOT TRAN V(1) V(1,2) V(4)
.PROBE
.END
```

FIGURE 19–1

120

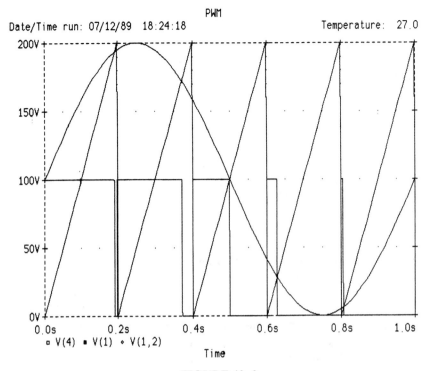

FIGURE 19–2

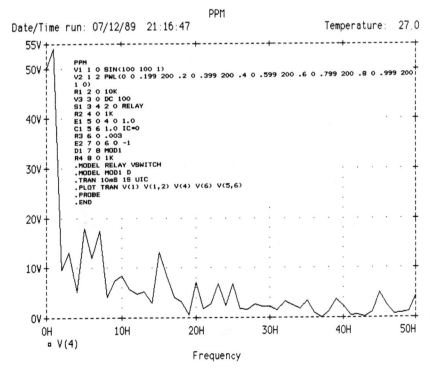

FIGURE 19–3

frequency components of the PWM waveform occur at up to 50 Hz. Even if the frequency were limited to components of less than 25 Hz, which would produce rounded corners and less accuracy in the PWM, the bandwidth required to pass PWM would still be 25 times the bandwidth of the analog input that is being sampled. This large-bandwidth requirement makes PWM unattractive for most communications applications.

PPM

Figure 19–1 shows how PPM can be derived from PWM by simple differentiation, inversion, and clipping. If the PWM pulses of Figure 19–2 are differentiated, the result is a series of short positive and negative pulses at the leading and the trailing edge of each PWM pulse. These pulses are shown in Figure 19–4 with the input analog sine wave included for reference. The negative pulses shown in Figure 19–4 occur at a distance from the beginning of each 0.2-s clock cycle that is proportional to the amplitude of the sinusoidal input. Thus, the negative pulses represent PPM. If a diode is used to clip the positive pulses in Figure 19–4 and the negative pulses are then inverted (inverting amplifier), the result is PPM (exponential pulses). The PPM pulses are shown in Figure 19–5 with the input sinusoid for reference. Note in Figure 19–5 that the pulses occur at a distance from the beginning of each clock pulse (0.2, 0.4, and so forth) that is directly proportional to the amplitude of the sinusoid. A one-shot multivibrator can be used (as shown in Figure 19–1) to produce PPM pulses with square corners. The frequency spectrum of the exponential PPM of Figure 19–5 is shown in Figure 19–6. As

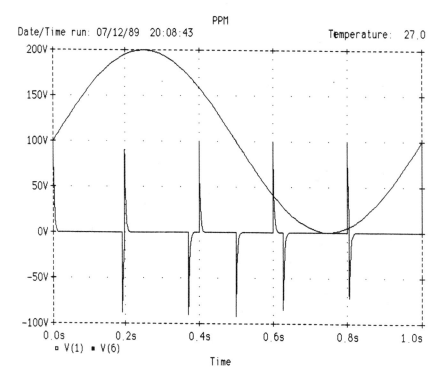

FIGURE 19–4

122

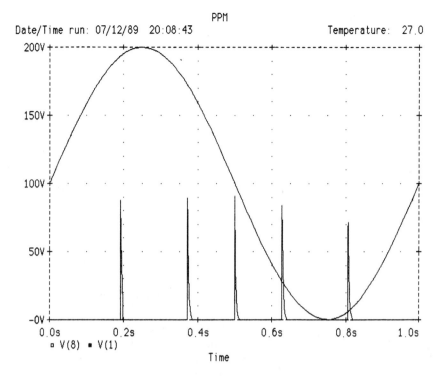

PPM

FIGURE 19–5

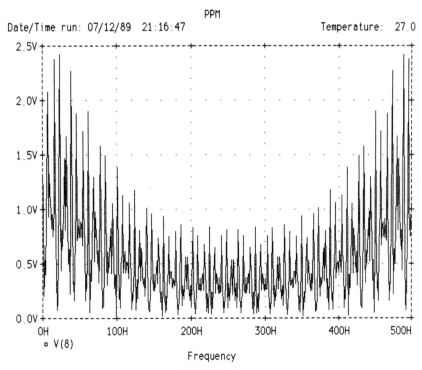

FIGURE 19–6

you can see, there are many frequency components for these short pulses. This means that a very wide bandwidth is required to carry the information in PPM. This wide-bandwidth requirement is the primary reason that PPM is not being widely used in communications.

Experimental Circuits

Figure 19–7 shows the LM555 circuit used to produce PWM in this experiment. The internal components of the 555 are shown in the diagram. From this it can be deduced that the input clock (a squarewave) sets the comparator (at pin 2) high at the beginning of each clock pulse. The comparator then sets the flip-flop high. The modulation into pin 5 is compared with a ramp (formed by the capacitor at pin 6 charging through the *npn* transistor), and the output resets the flip-flop when the ramp exceeds the input analog voltage. The 555 essentially acts like the comparator example from Figure 19–1. Adjustment of the potentiometer in Figure 19–7 sets the dc level of the ramp to get a change in width of the PWM output for the entire input waveform.

Figure 19–8 shows the LM555 circuit to produce PPM in this experiment. The trigger input occurs whenever the 0.033 μF capacitor charges to greater than one-half of the input voltage (because of the internal resistor network). This turns the flip-flop on. The flip-flop is turned off when the input modulation voltage exceeds the capacitor voltage. PPM is produced; however, these are not short pulses, but pulses that begin where the analog input exceeds the ramp and end essentially at the beginning of the next cycle. This waveform is PPM, but care must be taken in triggering and observation, since it can be mistaken for a PWM waveform, as will be seen in the experiment.

Demodulation

One method for demodulating PWM is shown in Figure 19–9. Because the length of the pulse in PWM is proportional to the amplitude of the analog input waveform and it is a

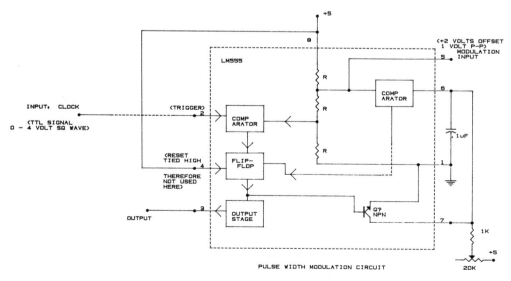

PULSE WIDTH MODULATION CIRCUIT

FIGURE 19–7

124

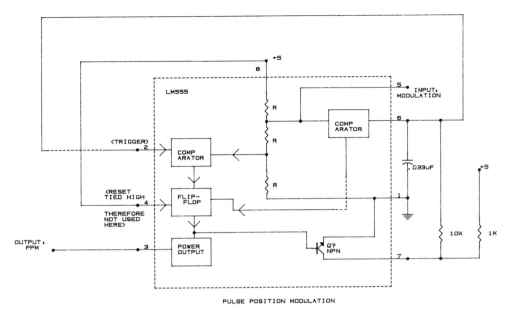

PULSE POSITION MODULATION

FIGURE 19–8

rectangular pulse, the area under the PWM waveform is also proportional to the amplitude of the input. Thus, a simple integrator circuit will provide a waveform that resembles the original analog input. Of course, smoothing must be done, but the frequency spectrum of the output of the integrator will contain the analog input frequencies as its principal components.

Demodulation of PPM can also be accomplished using several methods. One of the simplest is to convert the PPM waveform into a PWM waveform and then demodulate the PWM using the method just discussed. Figure 19–9 uses an RS flip-flop to convert the PPM input to PWM. The key to this method is that, as in phase modulation, the original clock input must be available. This makes demodulation of PPM more difficult (as well as more expensive) and along with the bandwidth requirements makes it an unattractive communications method in most instances.

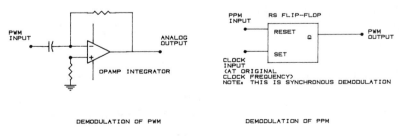

DEMODULATION OF PWM   DEMODULATION OF PPM

FIGURE 19–9

PROCEDURE AND QUESTIONS/PROBLEMS

Procedure

(NOTE: The main problem in this lab will be synchronizing the oscilloscope to observe the waveforms, especially PPM.)

PWM

1. Construct the PWM circuit of Figure 19–7.
2. Connect the input clock signal (a 0 to 4 V, 1600 Hz squarewave from a signal generator can simulate the TTL input). A 555 can also be used to generate the clock.
3. Connect the modulation input. Initially use a 1 V pk-pk, 300 Hz sinusoid with a +2 V offset.
4. Synchronize the oscilloscope using the input sinusoid. Adjust the potentiometer until the width of the PWM pulses is a maximum at the positive peak of the sinusoid and a minimum at the negative peak of the sinusoid. (Try inverting the PWM if the minimums and maximums are reversed). The minimum pulse width may be one-half of the period and the maximum almost the full width of the period if you are using a squarewave input. If the pulses drift to the right or left, change the *frequency* of the modulation input slightly until it is steady.
5. Observe and sketch your PWM waveform.
6. Change the input analog modulation to a triangular waveform and sketch the PWM output.
7. Explain the PWM outputs of parts 5 and 6.

PPM

8. Connect the PPM circuit shown in Figure 19–8.
9. Connect a triangular-wave modulation input. Use the same modulation amplitude, offset, and frequency that you used for PWM.
10. Synchronize the oscilloscope on the input waveform and observe the PPM output. Change the *frequency* of the input triangular wave slightly to stabilize the PPM output. This probably looks like PWM at this point. To verify that PPM is produced, compare the input triangular wave with the waveform at pin 6. Note that the fast ramp waveform (at the clock frequency), follows the slower input analog waveform, so that you have a fast ramp-type wave riding on a slower triangular wave. If you compare the start times of the PPM output, you will see that they begin at a point that is proportional to the amplitude of the input. The width of these PPM pulses has no bearing on their being PPM.
11. Observe and sketch your PPM waveform for the triangular-wave input. (This will also work with a sinusoidal input, but the triangular input produces a more readily recognizable output).

Questions/Problems

1. Draw and explain the circuit used for simulating the PWM waveform using PSPICE.
2. Draw and explain the circuit used for simulating the PPM waveform using PSPICE.

20

ADAPTIVE DELTA MODULATION
AND DEMODULATION

INTRODUCTION

Adaptive delta modulation is a simple, yet powerful, method to transmit voice band-width signals using digital methods. This technique is widely used by telephone companies and in the latest military VHF frequency-hopping tactical digital radios. The purpose of this experiment is to familiarize you with the theory and practice of adaptive delta modulation and to introduce a modern single-chip coder-decoder (CODEC). The MC3417 continuously variable slope delta (CVSD) modem is a CODEC that provides adaptive delta modulation (CVSD) and is converted from a receiver to a transmitter with a single ''push-to-talk'' switch. In the experiment you will observe the operation of a CVSD modem and use it to both transmit and receive sinusoid and voice signals.

After completing this experiment, you will be able to

1. Understand delta and adaptive delta (also known as CVSD) operation.
2. Construct a CVSD modem using the MC3417.

REFERENCES

1. Young, Chapter 11.
2. Roddy and Coolen, Chapter 17.
3. Miller, Chapter 9.
4. Tomasi (*Advanced*), Chapter 4.
5. Killen, Chapters 11 and 12.
6. Adamson, Chapter 8.

MATERIALS OR SPECIAL INSTRUMENTATION

Two function generators

Devices: MC3417 CVSD CODEC, SPST switch, radio with ear-jack output or microphone, audio amplifier, speaker

Resistors: four 1 k, 1.2 k, 3.3 k, four 10 k, 18 k, 2.4 MΩ

Capacitors: two 0.01 μF, two 1 μF, 0.22 μF, 4.7 μF

THEORETICAL BACKGROUND

Digital transmission of analog signals is important because digital signals can be completely regenerated (with no noise), digital circuits are relatively inexpensive and easy to work with, and computer manipulation of digital signals allows extensive signal processing. As with any method of transmission, the simpler the circuitry, the better. With what we think of as "normal" pulse-code modulation, the amplitude of the analog signal is sampled, and the samples are coded into 8, 10, 12, or 16 bits (depending on the accuracy required). This group of bits must then be transmitted along with protocol and error-detection information to represent a single sample of the analog signal. This method creates a great deal of overhead (complication), which makes the real-time transmission of information necessary for voice conversations more difficult.

Delta Modulation

Delta modulation provides a simpler, more direct method of transmitting voice information. In its simplest form, delta modulation transmits a single bit per sample of the analog waveform. Figure 20–1 shows a block diagram of the CVSD chip used in this experiment. Without the two blocks inside the dashed box in Figure 20–1, the system is a simple delta modulator. This delta modulator consists of the following:

1. A comparator, which compares the audio input with the (demodulated) analog output.
2. A sampler that samples the comparator output and thus provides either a high (1) if the audio input is greater or a low (0) if the demodulated output is higher.

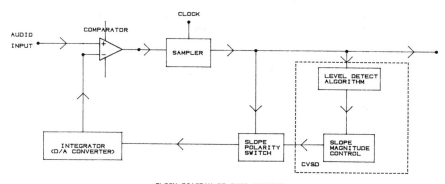

BLOCK DIAGRAM OF CVSD ENCODER
(FROM MC3417 SPECIFICATIONS, SEE APPENDIX A)

FIGURE 20–1

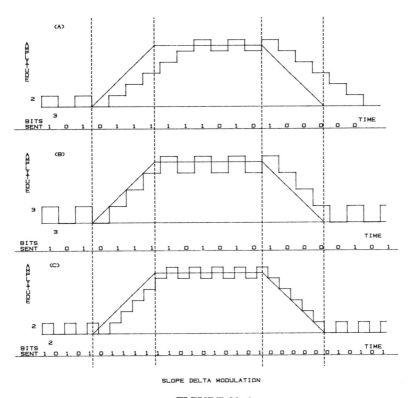

SLOPE DELTA MODULATION

FIGURE 20-2

3. A slope polarity switch that determines (based on the last several samples) if the slope of the audio input is positive or negative (increasing or decreasing in amplitude).

4. An integrator, or DAC, to add each successive bit to the total (if the slope polarity switch is positive) or subtract the bit (if the slope polarity switch is negative).

Note that the slope polarity switch and the integrator are all that are required to demodulate the basic slope delta modulation at the receiver and convert it back to an analog waveform.

Figure 20-2 illustrates basic delta modulation used on a trapezoidal waveform. In Figure 20-2A a fixed pulse width of three units and a fixed pulse height of two units are used. The bits sent on the communications channel are shown below the figure, and the output of the integrator is shown as the stair-step waveform with the trapezoidal analog input. The delta modulation follows the trapezoid to a certain extent, with two problems:

1. There is a delay when the slope of the analog signal changes (the sides of the trapezoid). This also indicates that the delta modulation cannot follow the slope of the analog input.

2. There is a lot of noise when a level (dc) is reached (at the top of the trapezoid).

129

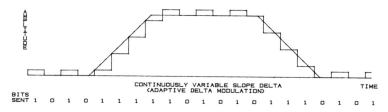

CONTINUOUSLY VARIABLE SLOPE DELTA
(ADAPTIVE DELTA MODULATION)   TIME

BITS
SENT 1 0 1 0 1 1 1 1 1 0 1 0 1 0 1 1 1 0 1 0 1

FIGURE 20–3

In Figure 20–2B each bit represents a larger step. Now, the delta modulation can follow the analog input when its slope changes, but there is a lot of noise at a dc level (more than in part A).

In Figure 20–2C a smaller bit length (faster clock) is chosen, with the amplitude represented by each bit, as in part A. The delta modulation can now follow the slope better than in A and there is less noise at a dc level than in B, but the price is a faster clock and a higher bit rate.

This illustrates a major problem with simple delta modulation: There is a trade-off between sampling rate and step size in order for the digital slope delta modulation to follow the analog waveform correctly. The result is a system that is tailored for one particular waveform and is not adaptable to something as variable as a voice signal.

CVSD

Adaptive delta modulation, or continuously variable slope delta (CVSD) as the MC3417 calls this technique, enables practical transmission of voice signals by varying the amplitude represented by each bit, depending on the slope of the input analog signal. Figure 20–3 shows a trapezoidal input with its bit output shown below the figure. The amplitude represented by each bit is initially 1 unit. As the slope of the trapezoid begins, the slope magnitude control (shown in the block diagram of Figure 20–1) is increased until the input levels off, and it is again decreased. The amplitude of the steps is also varied with a negative slope, as shown in Figure 20–3.

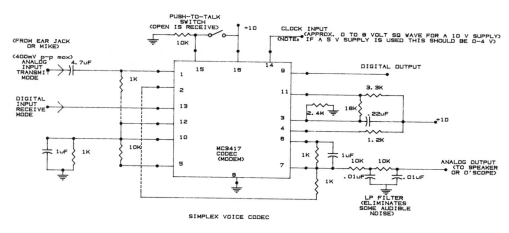

SIMPLEX VOICE CODEC

FIGURE 20–4

The key to this adaptive delta modulation is a digital algorithm that enables the demodulator (slope polarity switch and integrator in Figure 20–1) to know when to increase the integration constant. The MC3417 uses a 3-bit algorithm that merely keeps the last three output bits in a shift register. If all three bits are 1s, it means the digital output is not keeping up with the analog input, and the slope is automatically increased. If all three bits are 0s, it means that the slope is too large, and it is decreased. Note that this algorithm is not used in Figure 20–3 due to space limitations, but it works quite well. The MC3418 uses a 4-bit algorithm but requires a generally faster clock rate.

An excellent explanation of the internal operation of the MC3417 is contained in the specifications in the Appendix. Figure 20–4 shows the CVSD circuit to be used in this experiment.

PROCEDURE AND QUESTIONS/PROBLEMS

Procedure

1. Construct the circuit of Figure 20–4.
2. Use a function generator (0- to 8-V squarewave) at 60 kHz for the clock (of course, a 50 percent duty cycle 555 will work well) and another function generator for the analog input at an audio frequency (approximately 400 mV pk-pk), and compare the input (pin 1) and the reconstructed analog output (pin 7). The waveform at pin 7 is the integrated digital signal and is difficult to synchronize with an analog oscilloscope. This waveform (at pin 7) should be similar to that shown in Figure 12 of the MC3417 specifications in the Appendix. Sketch your output and comment on the operation of the circuit.
3. Compare the digital output of the circuit (pin 9) with the analog input. Analyze the digital output over one period of the sine wave and use this to illustrate how the CVSD algorithm operates.
4. Connect your CODEC to that of another group with a wire (preferably across the room), using the same clock frequency for both CODECs. Two different signal generators will work, but best reception occurs when both have the same clock frequency, so use a frequency counter to ensure this. Attach a radio ear jack or microphone to the input of each circuit (you can initially test reception of an input sinusoid) and a speaker to the output of each circuit. Test the digital transmission of voice signals in both directions in the simplex mode. Note that an FM station will provide a wider bandwidth signal and, hence, better audible results. Comment on your results.
5. Slowly change one clock frequency and comment on the resulting loss of signal (and resulting noise) as the difference between the two clock frequencies varies.
6. Test the setup in both directions. Then reduce both clock frequencies to 16 kHz (frequency counter) and comment on the results.

Questions/Problems

1. Explain the operation of the simplex circuit you and the other lab group combined to build.
2. What is the function of a syllabic filter?

21

FIBER-OPTIC PRINCIPLES

INTRODUCTION

Communications using fiber-optic (FO) cable (waveguide) use time-division multiplexing (TDM) and transmit information in digital form. FO cables are used by major telephone companies, utility companies, and all industries that have a need for wide-bandwidth (fast data rate) transmission of information with no electromagnetic interference. Modern FO cables are lightweight and relatively low in cost and have an enormous bandwidth capability. The purpose of this experiment is to introduce some basic principles of fiber-optic transmission and reception and to delineate characteristics of the FO waveguides and optical transmitters and receivers. The devices used here can be simple (such as the MFOE71/MFOD72 transmitter/receiver and 1000 μm plastic FO cable shown in Figure 21–1) or more sophisticated, depending on available equipment and material.

After completing this experiment, you will be able to

1. Understand the basic physical principles that set the primary limits on fiber-optic bit rate.
2. Build a basic FO transmitter and receiver.
3. Design a basic TDM FO transmitter and receiver.

REFERENCES

1. Young, Chapter 18.
2. Roddy and Coolen, Chapter 20.
3. Miller, Chapter 15.

4. Tomasi (*Advanced*), Chapter 10.

5. Killen, Chapter 14.

6. Adamson, Chapter 12.

MATERIALS OR SPECIAL INSTRUMENTATION

Devices: MFOE71 (Motorola or Radio Shack) infrared-emitting diode (or other IR emitter), MFOD72 (Motorola or Radio Shack) infrared detector (or other IR detector), two 2N3904 *npn* transistors, optical waveguide (to fit the emitter and detector), TTL inverter (or 7400 NAND gate), TTL XOR gate (7486)

Resistors: 33 k, 3.9 k, three 1 k

THEORETICAL BACKGROUND

General

Information transmitted using FO waveguides is generally encoded digitally, and separate information channels are combined on a single FO cable using TDM. Frequency-division multiplexing (FDM) is not currently used because the sources of IR light (either a laser or an IR-emitting diode) have a relatively wide bandwidth. For instance, at an IR wavelength of 1.3 μm (= 1300 nm = 13,000 Å), the frequency is f = (speed of light)/ (wavelength) = $3 \times 10^8/1.3 \times 10^{-6} = 2.3 \times 10^{14}$, or 230,000 GHz. If the IR emitter is a relatively narrow bandwidth solid-state laser with a 1 percent BW, the BW of the emitter is still 2300 GHz. This, along with the difficulty of detection, does not allow FDM to be used commercially at the present time.

Pulse Spreading

The enormous bandwidth of light-wave communications illustrated in the preceding calculation makes it attractive in bandwidth-limited areas of communications, with commercial telephone systems being the most widely publicized. Perhaps the most fundamental limitation on the bit rate that can be transmitted over a FO cable is *pulse spreading*. This pulse spreading can be caused by the rise time or fall time of the transmit and receive devices (as you will see in the experiment); however, for commercial, high-bit-rate lines with high-speed (less than 1 ns rise and fall time) transmitters and receivers, the dispersion in the cable itself is the limiting factor. One type of dispersion is the difference in arrival time of pulses due to different path lengths. This is the only significant dispersion mechanism in *step-index fibers* and is still the major source of dispersion in *graded-index fibers*. Note that step-index fibers are not currently used much commercially due to narrow bandwidth and that graded-index fibers are currently the primary commercial FO cable because of cost/bandwidth considerations.

The third type of fiber is *single-mode fiber*, which does not have any path-length dispersion. Material dispersion (dispersion due to material characteristics) is the primary mechanism for pulse spreading in single-mode fibers. Single-mode fibers have the largest bandwidth and bit rate, but the cable is expensive (it is very small, less than 10 μm in diameter, and it must be manufactured with very tight tolerances). Single-mode

134

fibers are widely used for systems with extremely wide bandwidth requirements, such as undersea cables.

Path-length dispersion and material dispersion both cause a difference in arrival time of a pulse (bit) at the receiver. If the path length is different for different modes in a FO waveguide, and the speed of light in the material is the same, path-length dispersion occurs. For instance, a difference in path length of 1 m (in, say, a 20-km cable) with a velocity in the material of two-thirds the speed of light will cause a difference in arrival time of $1 \text{ m}/2 \times 10^8 \text{ m/s} = 0.5 \times 10^{-8}$ s, or 5 ns. This will spread the arrival time of a pulse by 5 ns and means that a maximum bit rate of $1/5 \times 10^{-9} = 200$ MHz is allowable, even with very short transmission pulses.

Material dispersion is caused by a variance in the speed of light in the material at different frequencies. Because the LED or solid-state laser transmitters have relatively wide bandwidths, there is a considerable variation in the transmission frequency when one is turned on and off (which is how 1s and 0s are sent). This frequency variation at the input combined with the variation in the speed of light at different frequencies due to the material causes a pulse spreading that creates the same effect at the receiver as the previously discussed path-length dispersion.

A Transmitter and Receiver to Illustrate Pulse Spreading

Figure 21–1 shows a basic FO link to transmit TTL information and to observe the pulse-spreading phenomena at different frequencies. The TTL input turns the 2N3904 transistor on and off, which sends a pulse of current through the IR-emitting diode with each TTL 1. Thus, a pulse of IR energy (photons) is emitted from the transmitter into the FO cable with each 1. The detector diode is turned on when it receives sufficient IR energy (photons) from the cable. Because the received energy is relatively small, and hence the current output of the receiver is small, a second 2N3904 is used to amplify the output current. This causes an inversion in the TTL waveform, which is reinverted with the NAND gate (or with an inverter). The TTL output as shown should be the same as the TTL input except for the pulse-spreading effect.

The XOR gate (7486) shown is used to compare the input TTL and the output TTL waveforms. The output of the XOR gate is, thus, a pulse train of varying width that

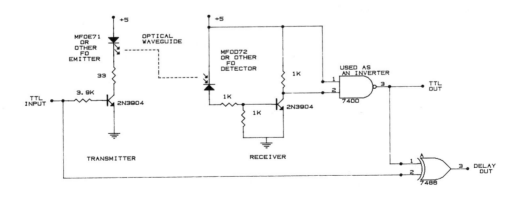

A SIMPLE TRANSMITTER & RECEIVER TO SHOW PULSE SPREADING

FIGURE 21–1

shows the delay (pulse spreading) from the input TTL to the output of the system. This, of course, can also be accomplished directly on most oscilloscopes. As the frequency of the input TTL squarewave is increased, a frequency is reached at which the output delay is longer than the input pulse; this marks the highest frequency that can be transmitted with this transmitter/receiver.

Even a multimode fiber will generally have a maximum frequency of many megahertz over a very short path length. This would be difficult to observe without special equipment, so to get an easily observed waveform, the delay in this setup is due to characteristics of the transmitter and the receiver. From the specifications (see Appendix), the optical rise and fall time for the transmitter is 25 ns, so this would provide a delay (pulse spreading) of about 25 ns and a maximum frequency of transmission of about 1/25 ns = 40 Mrad/s, or 6.36 MHz. Next, the turn-on and turn-off times for the detector are listed (see Appendix) as 10 μs and 60 μs, respectively. This is quite slow (compared with the 25 ns transmitter) and will provide the limit for the TTL bit rate. For a simple estimate, inverting the 60 μs gives a frequency of about 16.7 krad/s, or 2.7 kHz as a maximum usable frequency. Since you can theoretically send 2 bits/Hz, the maximum bit rate would be about 5.3 kbits/s. Since this is based on a specification (worst case), you expect to be able to transmit and receive a greater bit rate than this.

TDM Applied to Fiber-Optics

Application of the TDM circuit of Experiment 9 to this FO circuit is quite simple. Up to four inputs (TTL), each with a low bit-rate, can be directly time-division multiplexed using Figure 9–1. Timing is critical when sending groups of bits, and the 555 timer in Figure 9–1 would have to ensure that an entire group of bits is sent. Many protocols have been developed to ensure proper signal transmission and reception on local area networks (LANs), and the like, so this will not be discussed further.

The FO link can still be used to transmit and receive a TDM analog signal; however, the emitter diode must be biased at a point that is barely off (a small current with very little IR output) rather than all the way off (zero current and zero IR output), so that the analog signal is not cut off during part of the signal. This requires a biasing network for the input and output 2N3904s and elimination of the TTL gate. If this is done, the output of the 4052 in Figure 9–1 can be connected directly to the input of the FO circuit and the signal transmitted over the FO waveguide. Note that the 1-kHz sampling signal of Figure 9–1 still requires that the maximum frequency of the transmitted signal be less than 500 Hz because of Nyquist's sampling theorem.

PROCEDURE AND QUESTIONS/PROBLEMS

Procedure

1. Construct the circuit of Figure 21–1.
2. Using a squarewave input from a signal generator, at TTL levels (about 0 to 4 V), determine the maximum data rate (remember, you are sending 2 bits/cycle with a square wave) in bits per second that could be transmitted over your circuit.

Questions/Problems

1. Discuss your results and what needs to be done to transmit and receive a higher bit-rate signal.
2. Design a block diagram of a TDM system for transmission of four 1200-bit/s signals using the basic TDM system of Figure 9–1 and the FO system of Figure 21–1.

APPENDIX

MANUFACTURERS' SPECIFICATIONS

National Semiconductor

LM741/LM741A/LM741C/LM741E Operational Amplifier

General Description

The LM741 series are general purpose operational amplifiers which feature improved performance over industry standards like the LM709. They are direct, plug-in replacements for the 709C, LM201, MC1439 and 748 in most applications.

The amplifiers offer many features which make their application nearly foolproof: overload protection on the input and output, no latch-up when the common mode range is exceeded, as well as freedom from oscillations.

The LM741C/LM741E are identical to the LM741/LM741A except that the LM741C/LM741E have their performance guaranteed over a 0°C to +70°C temperature range, instead of −55°C to +125°C.

Schematic and Connection Diagrams (Top Views)

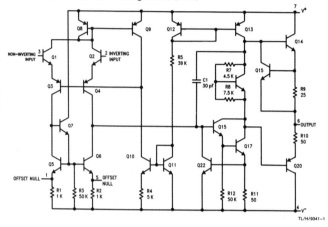

TL/H/9341–1

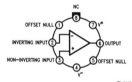

Metal Can Package

TL/H/9341–2

Order Number LM741H, LM741AH,
LM741CH or LM741EH
See NS Package Number H08C

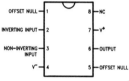

Dual-In-Line or S.O. Package

TL/H/9341–3

Order Number LM741J, LM741AJ, LM741CJ,
LM741CM, LM741CN or LM741EN
See NS Package Number J08A, M08A or N08E

LIFE SUPPORT POLICY

NATIONAL'S PRODUCTS ARE NOT AUTHORIZED FOR USE AS CRITICAL COMPONENTS IN LIFE SUPPORT DEVICES OR SYSTEMS WITHOUT THE EXPRESS WRITTEN APPROVAL OF THE PRESIDENT OF NATIONAL SEMICONDUCTOR CORPORATION. As used herein:

1. Life support devices or systems are devices or systems which, (a) are intended for surgical implant into the body, or (b) support or sustain life, and whose failure to perform, when properly used in accordance with instructions for use provided in the labeling, can be reasonably expected to result in a significant injury to the user.

2. A critical component is any component of a life support device or system whose failure to perform can be reasonably expected to cause the failure of the life support device or system, or to affect its safety or effectiveness.

National Semiconductor Corporation
2900 Semiconductor Drive
P.O. Box 58090
Santa Clara, CA 95052-8090
Tel: (408) 721-5000
TWX: (910) 339-9240

National Semiconductor GmbH
Westendstrasse 193-195
D-8000 Munchen 21
West Germany
Tel: (089) 5 70 95 01
Telex: 522772

NS Japan Ltd.
Sanseido Bldg, 5F
4-15 Nishi Shinjuku
Shinjuku-Ku,
Tokyo 160, Japan
Tel: 3-299-7001
FAX: 3-299-7000

National Semiconductor Hong Kong Ltd.
Southeast Asia Marketing
Austin Tower, 4th Floor
22-26A Austin Avenue
Tsimshatsui, Kowloon, H.K.
Tel: 3-7231290, 3-7243645
Cable: NSSEAMKTG
Telex: 52996 NSSEA HX

National Semicondutores Do Brasil Ltda.
Av. Brig. Faria Lima, 830
8 Andar
01452 Sao Paulo, SP, Brasil
Tel: (55/11) 212-5066
Telex: 391-1131931 NSBR BR

National Semiconductor (Australia) PTY, Ltd.
21/3 High Street
Bayswater, Victoria 3153
Australia
Tel: (03) 729-6333
Telex: AA32096

National does not assume any responsibility for use of any circuitry described, no circuit patent licenses are implied and National reserves the right at any time without notice to change said circuitry and specifications.

MPF102

CASE 29-02, STYLE 5
TO-92 (TO-226AA)

JFET
VHF AMPLIFIER

N-CHANNEL — DEPLETION

Refer to 2N4416 for graphs.

MAXIMUM RATINGS

Rating	Symbol	Value	Unit
Drain-Source Voltage	V_{DS}	25	Vdc
Drain-Gate Voltage	V_{DG}	25	Vdc
Gate-Source Voltage	V_{GS}	− 25	Vdc
Gate Current	I_G	10	mAdc
Total Device Dissipation @ T_A = 25°C Derate above 25°C	P_D	200 2	mW mW/°C
Junction Temperature Range	T_J	125	°C
Storage Temperature Range	T_{stg}	− 65 to + 150	°C

ELECTRICAL CHARACTERISTICS (T_A = 25°C unless otherwise noted.)

Characteristic	Symbol	Min	Max	Unit
OFF CHARACTERISTICS				
Gate-Source Breakdown Voltage (I_G = − 10 µAdc, V_{DS} = 0)	$V_{(BR)GSS}$	− 25	—	Vdc
Gate Reverse Current (V_{GS} = − 15 Vdc, V_{DS} = 0) (V_{GS} = − 15 Vdc, V_{DS} = 0, T_A = 100°C)	I_{GSS}	— —	− 2.0 − 2.0	nAdc µAdc
Gate Source Cutoff Voltage (V_{DS} = 15 Vdc, I_D = 2.0 nAdc)	$V_{GS(off)}$	—	− 8.0	Vdc
Gate Source Voltage (V_{DS} = 15 Vdc, I_D = 0.2 mAdc)	V_{GS}	− 0.5	− 7.5	Vdc
ON CHARACTERISTICS				
Zero-Gate-Voltage Drain Current* (V_{DS} = 15 Vdc, V_{GS} = 0 Vdc)	I_{DSS}	2.0	20	mAdc
SMALL-SIGNAL CHARACTERISTICS				
Forward Transfer Admittance* (V_{DS} = 15 Vdc, V_{GS} = 0, f = 1.0 kHz) (V_{DS} = 15 Vdc, V_{GS} = 0, f = 100 MHz)	$|y_{fs}|$	2000 1600	7500 —	µmhos
Input Admittance (V_{DS} = 15 Vdc, V_{GS} = 0, f = 100 MHz)	$Re(y_{is})$	—	800	µmhos
Output Conductance (V_{DS} = 15 Vdc, V_{GS} = 0, f = 100 MHz)	$Re(y_{os})$	—	200	µmhos
Input Capacitance (V_{DS} = 15 Vdc, V_{GS} = 0, f = 1.0 MHz)	C_{iss}	—	7.0	pF
Reverse Transfer Capacitance (V_{DS} = 15 Vdc, V_{GS} = 0, f = 1.0 MHz)	C_{rss}	—	3.0	pF

*Pulse Test: Pulse Width ≤ 630 ms; Duty Cycle ≤ 10%.

141

2N5457
2N5458
2N5459

CASE 29-02, STYLE 5
TO-92 (TO-226AA)

JFET
GENERAL PURPOSE

N-CHANNEL — DEPLETION

Refer to 2N4220 for graphs.

MAXIMUM RATINGS

Rating	Symbol	Value	Unit
Drain-Source Voltage	V_{DS}	25	Vdc
Drain-Gate Voltage	V_{DG}	25	Vdc
Reverse Gate-Source Voltage	V_{GSR}	−25	Vdc
Gate Current	I_G	10	mAdc
Total Device Dissipation @ T_A = 25°C Derate above 25°C	P_D	310 2.82	mW mW/°C
Junction Temperature Range	T_J	125	°C
Storage Channel Temperature Range	T_{stg}	−65 to +150	°C

ELECTRICAL CHARACTERISTICS (T_A = 25°C unless otherwise noted.)

Characteristic		Symbol	Min	Typ	Max	Unit
OFF CHARACTERISTICS						
Gate-Source Breakdown Voltage (I_G = −10 μAdc, V_{DS} = 0)		$V_{(BR)GSS}$	−25	—	—	Vdc
Gate Reverse Current (V_{GS} = −15 Vdc, V_{DS} = 0) (V_{GS} = −15 Vdc, V_{DS} = 0, T_A = 100°C)		I_{GSS}	— —	— —	−1.0 −200	nAdc
Gate Source Cutoff Voltage (V_{DS} = 15 Vdc, I_D = 10 nAdc)	2N5457 2N5458 2N5459	$V_{GS(off)}$	−0.5 −1.0 −2.0	— — —	−6.0 −7.0 −8.0	Vdc
Gate Source Voltage (V_{DS} = 15 Vdc, I_D = 100 μAdc) (V_{DS} = 15 Vdc, I_D = 200 μAdc) (V_{DS} = 15 Vdc, I_D = 400 μAdc)	2N5457 2N5458 2N5459	V_{GS}	— — —	−2.5 −3.5 −4.5	— — —	Vdc
ON CHARACTERISTICS						
Zero-Gate-Voltage Drain Current* (V_{DS} = 15 Vdc, V_{GS} = 0)	2N5457 2N5458 2N5459	I_{DSS}	1.0 2.0 4.0	3.0 6.0 9.0	5.0 9.0 16	mAdc
SMALL-SIGNAL CHARACTERISTICS						
Forward Transfer Admittance Common Source* (V_{DS} = 15 Vdc, V_{GS} = 0, f = 1.0 kHz)	2N5457 2N5458 2N5459	$|Y_{fs}|$	1000 1500 2000	— — —	5000 5500 6000	μmhos
Output Admittance Common Source* (V_{DS} = 15 Vdc, V_{GS} = 0, f = 1.0 kHz)		$|Y_{os}|$	—	10	50	μmhos
Input Capacitance (V_{DS} = 15 Vdc, V_{GS} = 0, f = 1.0 MHz)		C_{iss}	—	4.5	7.0	pF
Reverse Transfer Capacitance (V_{DS} = 15 Vdc, V_{GS} = 0, f = 1.0 MHz)		C_{rss}	—	1.5	3.0	pF

*Pulse Test: Pulse Width ≤ 630 ms; Duty Cycle ≤ 10%.

2N3903
2N3904

CASE 29-02, STYLE 1
TO-92 (TO-226AA)

GENERAL PURPOSE TRANSISTOR

NPN SILICON

MAXIMUM RATINGS

Rating	Symbol	Value	Unit
Collector-Emitter Voltage	V_{CEO}	40	Vdc
Collector-Base Voltge	V_{CBO}	60	Vdc
Emitter-Base Voltage	V_{EBO}	6.0	Vdc
Collector Current — Continuous	I_C	200	mAdc
Total Device Dissipation @ T_A = 25°C Derate above 25°C	P_D	625 5	mW mW/°C
*Total Device Dissipation @ T_C = 25°C Derate above 25°C	P_D	1.5 12	Watts mW/°C
Operating and Storage Junction Temperature Range	T_J, T_{stg}	− 55 to + 150	°C

*THERMAL CHARACTERISTICS

Characteristic	Symbol	Max	Unit
Thermal Resistance, Junction to Case	$R_{\theta JC}$	83.3	°C/W
Thermal Resistance, Junction to Ambient	$R_{\theta JA}$	200	°C/W

*Indicates Data in addition to JEDEC Requirements.

ELECTRICAL CHARACTERISTICS (T_A = 25°C unless otherwise noted.)

Characteristic		Symbol	Min	Max	Unit
OFF CHARACTERISTICS					
Collector-Emitter Breakdown Voltage(1) (I_C = 1.0 mAdc, I_B = 0)		$V_{(BR)CEO}$	40	—	Vdc
Collector-Base Breakdown Voltage (I_C = 10 μAdc, I_E = 0)		$V_{(BR)CBO}$	60	—	Vdc
Emitter-Base Breakdown Voltage (I_E = 10 μAdc, I_C = 0)		$V_{(BR)EBO}$	6.0	—	Vdc
Base Cutoff Current (V_{CE} = 30 Vdc, V_{EB} = 3.0 Vdc)		I_{BL}	—	50	nAdc
Collector Cutoff Current (V_{CE} = 30 Vdc, V_{EB} = 3.0 Vdc)		I_{CEX}	—	50	nAdc
ON CHARACTERISTICS					
DC Current Gain(1) (I_C = 0.1 mAdc, V_{CE} = 1.0 Vdc)	2N3903 2N3904	h_{FE}	20 40	— —	—
(I_C = 1.0 mAdc, V_{CE} = 1.0 Vdc)	2N3903 2N3904		35 70	— —	
(I_C = 10 mAdc, V_{CE} = 1.0 Vdc)	2N3903 2N3904		50 100	150 300	
(I_C = 50 mAdc, V_{CE} = 1.0 Vdc)	2N3903 2N3904		30 60	— —	
(I_C = 100 mAdc, V_{CE} = 1.0 Vdc)	2N3903 2N3904		15 30	— —	
Collector-Emitter Saturation Voltage(1) (I_C = 10 mAdc, I_B = 1.0 mAdc) (I_C = 50 mAdc, I_B = 5.0 mAdc)		$V_{CE(sat)}$	— —	0.2 0.3	Vdc
Base-Emitter Saturation Voltage(1) (I_C = 10 mAdc, I_B = 1.0 mAdc) (I_C = 50 mAdc, I_B = 5.0 mAdc)		$V_{BE(sat)}$	0.65 —	0.85 0.95	Vdc
SMALL-SIGNAL CHARACTERISTICS					
Current-Gain — Bandwidth Product (I_C = 10 mAdc, V_{CE} = 20 Vdc, f = 100 MHz)	2N3903 2N3904	f_T	250 300	— —	MHz

Signetics

MC1496/MC1596
Balanced Modulator/ Demodulator

Product Specification

Linear Products

DESCRIPTION
The MC1496 is a monolithic double-balanced modulator/demodulator designed for use where the output voltage is a product of an input voltage (signal) and a switched function (carrier). The MC1596 will operate over the full military temperature range of −55°C to +125°C. The MC1496 is intended for applications within the range of 0°C to +70°C.

FEATURES
● **Excellent carrier suppression**
  65dB typ @ 0.5MHz
  50dB typ @ 10MHz
● **Adjustable gain and signal handling**
● **Balanced inputs and outputs**
● **High common-mode rejection — 85dB typ**

APPLICATIONS
● **Suppressed carrier and amplitude modulation**
● **Synchronous detection**
● **FM detection**
● **Phase detection**
● **Sampling**
● **Single sideband**
● **Frequency doubling**

PIN CONFIGURATION

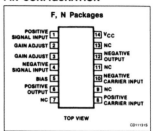

F, N Packages

Pin	Name	Pin	Name
1	POSITIVE SIGNAL INPUT	14	V_{CC}
2	GAIN ADJUST	13	NC
3	GAIN ADJUST	12	NEGATIVE OUTPUT
4	NEGATIVE SIGNAL INPUT	11	NC
5	BIAS	10	NEGATIVE CARRIER INPUT
6	POSITIVE OUTPUT	9	NC
7	NC	8	POSITIVE CARRIER INPUT

TOP VIEW

ORDERING INFORMATION

DESCRIPTION	TEMPERATURE RANGE	ORDER CODE
14-Pin Cerdip	0 to +70°C	MC1496F
14-Pin Plastic	0 to +70°C	MC1496N
14-Pin Cerdip	−55°C to +125°C	MC1596F
14-Pin Plastic	−55°C to +125°C	MC1596N

EQUIVALENT SCHEMATIC

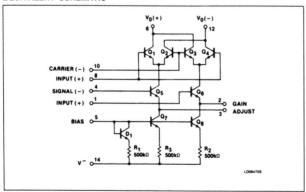

144

ABSOLUTE MAXIMUM RATINGS

SYMBOL	PARAMETER	RATING	UNIT
	Applied voltage	30	V
$V_8 - V_{10}$	Differential input signal	± 5.0	V
$V_4 - V_1$	Differential input signal	$(5 \pm I_5\, R_e)$	V
$V_2 - V_1$, $V_3 - V_4$	Input signal	5.0	V
I_5	Bias current	10	mA
P_D	Maximum power dissipation, $T_A = 25°C$ (still-air)[1] F package N package	1190 1420	mW mW
T_A	Operating temperature range MC1496 MC1596	0 to +70 −55 to +125	°C °C
T_{STG}	Storage temperature range	−65 to +150	°C

NOTE:
1. Derate above 25°C, at the following rates:
   F package at 9.5mW/°C.
   N package at 11.4mW/°C.

DC ELECTRICAL CHARACTERISTICS $V_{CC} = +12V_{DC}$; $V_{CC} = -8.0V_{DC}$; $I_5 = 1.0mA_{DC}$; $R_L = 3.9k\Omega$; $R_E = 1.0k\Omega$; $T_A = 25°C$, unless otherwise specified.

SYMBOL	PARAMETER	TEST CONDITIONS	MC1596			MC1496			UNIT
			Min	Typ	Max	Min	Typ	Max	
R_{IP} C_{IP}	Single-ended input impedance Parallel input resistance Parallel input capacitance	Signal port, f = 5.0MHz		200 2.0			200 2.0		$k\Omega$ pF
R_{OP} C_{OP}	Single-ended output impedance Parallel output resistance Parallel output capacitance	f = 10MHz		40 5.0			40 5.0		$k\Omega$ pF
I_{BS} I_{BC}	Input bias current $I_{BS} = \dfrac{I_1 + I_4}{2}$ $I_{BC} = \dfrac{I_8 + I_{10}}{2}$			12 12	25 25		12 12	30 30	μA μA
I_{IOS} I_{IOC}	Input offset current $I_{IOS} = I_1 - I_4$ $I_{IOC} = I_8 - I_{10}$			0.7 0.7	5.0 5.0		0.7 0.7	7.0 7.0	μA μA
$T_C I_{IO}$ I_{OO}	Average temperature coefficient of input offset current Output offset current $I_6 - I_{12}$			2.0 14	50		2.0 15	80	nA/°C μA
$T_C I_{OO}$ V_O	Average temperature coefficient of output offset current Common-mode quiescent output voltage (Pin 6 or Pin 12)			90 8.0			90 8.0		nA/°C V_{DC}
I_{D+} I_{D-}	Power supply current $I_6 + I_{12}$ I_{14}			2.0 3.0	3.0 4.0		2.0 3.0	4.0 5.0	mA_{DC}
P_D	DC power dissipation			33			33		mW

Copyright of Motorola, Inc. Used by Permission.

AC ELECTRICAL CHARACTERISTICS V_{CC} = +12$_{DC}$; V_{CC} = −9.0V_{DC}; I_5 = 1.0mA$_{DC}$; R_L = 3.9kΩ; R_E = 1.0kΩ; T_A = +25°C, unless otherwise specified.

SYMBOL	PARAMETER	TEST CONDITIONS	MC1596			MC1496			UNIT
			Min	Typ	Max	Min	Typ	Max	
V_{CFT}	Carrier feedthrough	V_C = 60mV$_{RMS}$ sinewave and offset adjusted to zero							
		f_C = 1.0kHz		40			40		μV_{RMS}
		f_C = 10MHz		140			140		
		V_C = 300mV$_{P-P}$ squarewave:							
		Offset adjusted to zero							
		f_C = 1.0kHz		0.04	0.2		0.04	0.4	mV$_{RMS}$
		Offset not adjusted f_C = 1.0kHz		20	100		20	200	
V_{CS}	Carrier suppressions	f_S = 10kHz, 300mV$_{RMS}$ sinewave	50	65		40	65		dB
		f_C = 500kHz, 60mV$_{RMS}$ sinewave							
		f_C = 10MHz, 60mV$_{RMS}$ sinewave		50			50		
BW$_{3dB}$	Transadmittance bandwidth (Magnitude) (R_L = 50Ω)	Carrier input port, V_C = 60mV$_{RMS}$ sinewave f_S = 1.0kHz, 300mV$_{RMS}$ sinewave		300			300		MHz
		Signal input port, V_S = 300mV$_{RMS}$ sinewave $|V_C|$ = 0.5V_{DC}		80			80		MHz
A_{VS}	Signal gain	V_S = 100mV$_{RMS}$; f = 1.0kHz $|V_C|$ = 0.5V_{DC}	2.5	3.5		2.5	3.5		V/V
CMV	Common-mode input swing	Signal port, f_S = 1.0kHz		5.0			5.0		V_{P-P}
A_{CM}	Common-mode gain	Signal port, f_S = 1.0kHz $|V_C|$ = 0.5V_{DC}		−85			−85		dB
DV$_{OUT}$	Differential output voltage swing capability			8.0			8.0		V_{P-P}

ECG724
DIFFERENTIAL RF/IF AMPLIFIER

description The **ECG724** is a monolithic RF/IF amplifier intended for emitter-coupled (differential) or cascode amplifier operation from DC to 120 MHz in industrial and communications equipment.

features

- Controlled for input offset voltage, input offset current, and input bias current
- Balanced differential amplifier configuration with controlled constant-current source to provide unexcelled versatility
- Single- and dual-ended operation
- Operation from DC to 120 MHz
- Balanced-AGC capability
- Wide operating-current range.

applications

- RF and IF linear amplifiers, both differential and cascode
- Mixers
- Oscillators
- Converters in commercial FM
- DC, audio and sense amplifiers
- Limiting IF amplifiers
- Hybrid building block
- Emitter coupled switches

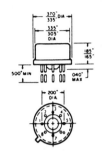

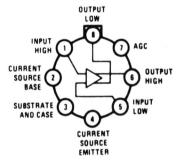

typical applications

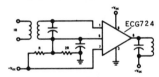

A Balanced Differential Amplifier with a Controlled Constant-Current-Source Drive and AGC Capability

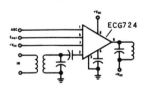

A Cascode Amplifier with a Constant-Impedance AGC Capability

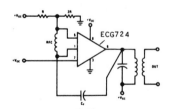

Oscillator

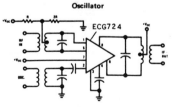

Mixer

dc electrical characteristics

	SYMBOL	TEST CIRCUIT	V_{CC}	V_{EE}	MIN	TYP	MAX	UNITS
Input Offset Voltage	V_{OS}	A	6	−6		0.4	2.0	mV
			12	−12		0.4	2.0	mV
Input Offset Current	I_{OS}	B	6	−6		0.15	2.0	µA
			12	−12		0.25	2.0	µA
Input Bias Current	I_{BIAS}	B	6	−6		7.5	40	µA
		B	12	−12		17	80	µA
Output Quiescent Operating Current	I_Q	B	6	−6	1.1	1.25	1.5	mA
		B	12	−12	2.5	3.15	4.0	mA
AGC Bias Current into Terminal 7	I_{AGC}	D	12	V_{AGC}=9V		1.1		mA
		D	12	V_{AGC}=12V		1.5		mA
			9					mA
			12					mA
Input Current into Terminal 7	I_7	B	6	−6	0.5	0.7	1.1	mA
		B	12	−12	1.0	1.5	2.2	mA
Power Dissipation	P_D	B	6	−6	24	35	42	mW
		B	12	−12	120	170	220	mW

ac electrical characteristics

	SYMBOL	TEST CIRCUIT	V_{CC}	V_{EE}	MIN	TYP	MAX	UNITS
100 MHz Power Gain	A_P	E(Cascode)	9	−	17	22		dB
		F(Diff.)	9	−	14.5	18.5		dB
10.7 MHz Power Gain	A_P	E(Cascode)	9	−	36	42		dB
		F(Diff.)	9	−	29	33.5		dB
100 MHz Noise Figure	NF	E(Cascode)	9	−		6.7	9.0	dB
		F(Diff.)	9	−		5.9	9.0	dB
Input Admittance at 10.7 MHz	Y_{11}	Cascode	+9	−		0.5+j1.3		mmho
		Diff.	+9	−		0.4+j0.58		mmho
Reverse Transadmittance at 10.7 MHz	Y_{12}	Cascode	+9	−		0.2+j0		µmho
		Diff.	+9	−		10+j0.2		µmho
Forward Transadmittance at 10.7 MHz	Y_{21}	Cascode	+9	−		95−j27		mmho
		Diff.	+9	−		−32+j.5		mmho
Output Admittance at 10.7 MHz	Y_{22}	Cascode	+9	−		0+j100		µmho
		Diff.	+9	−		20+j160		µmho
Output Power (Untuned) at 10.7 MHz	P_o	G	+9	−		5.7		µW
AGC Range at 10.7 MHz		F	+9	−		76		dB
Voltage Gain at 10.7 MHz	A_v	H(Cascode)	+9	−		40		dB
		I(Diff.)	+9	−		30		dB
Differential 1 kHz Voltage Gain	A_{vD}	J	6	−6	35	38	42	dB
		J	12	−12	40	42.5	45	dB
Maximum Peak to Peak Output Voltage at 1 kHz	$V_{OUT,pp}^{MAX}$	J R_L=2k	6	−6	8	11		V_{p-p}
		J R_L=1.6k	12	−12	16	22		V_{p-p}
3 dB Bandwidth	BW	J R_L=2k	6	−6		11.2		MHz
		J R_L=1.6k	12	−12		12.7		MHz
Common-Mode Input Voltage Range	V_{CM}	K	6	−6	−2.5	−3.2 to +4.5	4	V
		K	12	−12	−5	−7 to +9	7	V
Common-Mode Rejection Ratio	CMRR	K	6	−6	60	110		dB
		K	12	−12	60	90		dB
Peak to Peak Output Current V_{IN} = 400 mV at 10.7 MHz	I_{p-p}	Diff.	9	−	2.5	4.7	6	mA
		Diff.	12	−	4.5	6.5	8	mA

National Semiconductor

CD4051BM/CD4051BC Single 8-Channel Analog Multiplexer/Demultiplexer
CD4052BM/CD4052BC Dual 4-Channel Analog Multiplexer/Demultiplexer
CD4053BM/CD4053BC Triple 2-Channel Analog Multiplexer/Demultiplexer

General Description

These analog multiplexers/demultiplexers are digitally controlled analog switches having low "ON" impedance and very low "OFF" leakage currents. Control of analog signals up to $15V_{p-p}$ can be achieved by digital signal amplitudes of 3–15V. For example, if $V_{DD} = 5V$, $V_{SS} = 0V$ and $V_{EE} = -5V$, analog signals from $-5V$ to $+5V$ can be controlled by digital inputs of 0–5V. The multiplexer circuits dissipate extremely low quiescent power over the full $V_{DD} - V_{SS}$ and $V_{DD} - V_{EE}$ supply voltage ranges, independent of the logic state of the control signals. When a logical "1" is present at the inhibit input terminal all channels are "OFF".

CD4051BM/CD4051BC is a single 8-channel multiplexer having three binary control inputs. A, B, and C, and an inhibit input. The three binary signals select 1 of 8 channels to be turned "ON" and connect the input to the output.

CD4052BM/CD4052BC is a differential 4-channel multiplexer having two binary control inputs, A and B, and an inhibit input. The two binary input signals select 1 or 4 pairs of channels to be turned on and connect the differential analog inputs to the differential outputs.

CD4053BM/CD4053BC is a triple 2-channel multiplexer having three separate digital control inputs, A, B, and C, and an inhibit input. Each control input selects one of a pair of channels which are connected in a single-pole double-throw configuration.

Features

- Wide range of digital and analog signal levels: digital 3–15V, analog to $15V_{p-p}$
- Low "ON" resistance: 80Ω (typ.) over entire $15V_{p-p}$ signal-input range for $V_{DD} - V_{EE} = 15V$
- High "OFF" resistance: channel leakage of ± 10 pA (typ.) at $V_{DD} - V_{EE} = 10V$
- Logic level conversion for digital addressing signals of 3–15V ($V_{DD} - V_{SS} = 3$–15V) to switch analog signals to 15 V_{p-p} ($V_{DD} - V_{EE} = 15V$)
- Matched switch characteristics: $\Delta R_{ON} = 5\Omega$ (typ.) for $V_{DD} - V_{EE} = 15V$
- Very low quiescent power dissipation under all digital-control input and supply conditions: 1 μW (typ.) at $V_{DD} - V_{SS} = V_{DD} - V_{EE} = 10V$
- Binary address decoding on chip

Connection Diagrams

Dual-In-Line Packages

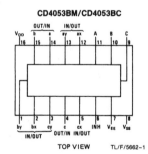

CD4051BM/CD4051BC

TOP VIEW

CD4052BM/CD4052BC

TOP VIEW

CD4053BM/CD4053BC

TOP VIEW

TL/F/5662–1

Order Number CD4051B*, CD4052B*, or CD4053B*

*Please look into Section 8, Appendix D for availability of various package types.

See the CMOS Logic Databook for Complete Specifications

149

CD4051BM/CD4051BC, CD4052BM/CD4052BC, CD4053BM/CD4053BC

DC Electrical Characteristics (Note 2) (Continued)

Symbol	Parameter	Conditions		−40°C Min	−40°C Max	+25°C Min	+25°C Typ	+25°C Max	+85°C Min	+85°C Max	Units
I_{DD}	Quiescent Device Current	$V_{DD} = 5V$			20			20		150	μA
		$V_{DD} = 10V$			40			40		300	μA
		$V_{DD} = 15V$			80			80		600	μA
Signal Inputs (V_{IS}) and Outputs (V_{OS})											
R_{ON}	"ON" Resistance (Peak for $V_{EE} \leq V_{IS} \leq V_{DD}$)	$R_L = 10\,k\Omega$ (any channel selected)	$V_{DD} = 2.5V$, $V_{EE} = -2.5V$ or $V_{DD} = 5V$, $V_{EE} = 0V$		850		270	1050		1200	Ω
			$V_{DD} = 5V$, $V_{EE} = -5V$ or $V_{DD} = 10V$, $V_{EE} = 0V$		330		120	400		520	Ω
			$V_{DD} = 7.5V$, $V_{EE} = -7.5V$ or $V_{DD} = 15V$, $V_{EE} = 0V$		210		80	240		300	Ω
ΔR_{ON}	Δ"ON" Resistance Between Any Two Channels	$R_L = 10\,k\Omega$ (any channel selected)	$V_{DD} = 2.5V$, $V_{EE} = -2.5V$ or $V_{DD} = 5V$, $V_{EE} = 0V$				10				Ω
			$V_{DD} = 5V$ $V_{EE} = -5V$ or $V_{DD} = 10V$, $V_{EE} = 0V$				10				Ω
			$V_{DD} = 7.5V$, $V_{EE} = -7.5V$ or $V_{DD} = 15V$, $V_{EE} = 0V$				5				Ω
	"OFF" Channel Leakage Current, any channel "OFF"	$V_{DD} = 7.5V$, $V_{EE} = -7.5V$ $O/I = \pm 7.5V$, $I/O = 0V$			±50		±0.01	±50		±500	nA
	"OFF" Channel Leakage Current, all channels "OFF" (Common OUT/IN)	Inhibit = 7.5V $V_{DD} = 7.5V$, $V_{EE} = -7.5V$ $O/I = 0V$ $I/O = \pm 7.5V$	CD4051		±200		±0.08	±200		±2000	nA
			CD4052		±200		±0.04	±200		±2000	nA
			CD4053		±200		±0.02	±200		±2000	nA
Control Inputs A, B, C and Inhibit											
V_{IL}	Low Level Input Voltage	$V_{EE} = V_{SS}$ $R_L = 1\,k\Omega$ to V_{SS} $I_{IS} < 2\,\mu A$ on all OFF Channels $V_{IS} = V_{DD}$ thru 1 kΩ $V_{DD} = 5V$			1.5			1.5		1.5	V
		$V_{DD} = 10V$			3.0			3.0		3.0	V
		$V_{DD} = 15V$			4.0			4.0		4.0	V
V_{IH}	High Level Input Voltage	$V_{DD} = 5$		3.5		3.5			3.5		V
		$V_{DD} = 10$		7		7			7		V
		$V_{DD} = 15$		11		11			11		V
I_{IN}	Input Current	$V_{DD} = 15V$, $V_{EE} = 0V$ $V_{IN} = 0V$			−0.1		-10^{-5}	−0.1		−1.0	μA
		$V_{DD} = 15V$, $V_{EE} = 0V$ $V_{IN} = 15V$			0.1		10^{-5}	0.1		1.0	μA

Note 1: "Absolute Maximum Ratings" are those values beyond which the safety of the device cannot be guaranteed. Except for "Operating Temperature Range" they are not meant to imply that the devices should be operated at these limits. The table of "Electrical Characteristics" provides conditions for actual device operation.

Note 2: All voltages measured with respect to V_{SS} unless otherwise specified.

150

Block Diagrams

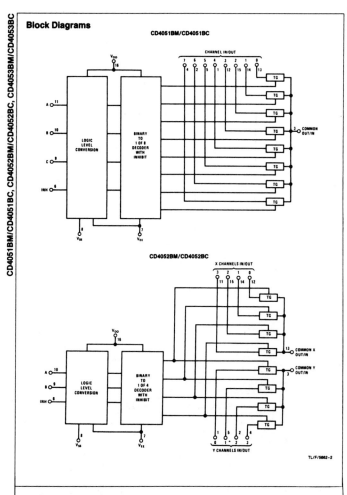

CD4051BM/CD4051BC

CD4052BM/CD4052BC

TL/F/5662-2

LIFE SUPPORT POLICY

NATIONAL'S PRODUCTS ARE NOT AUTHORIZED FOR USE AS CRITICAL COMPONENTS IN LIFE SUPPORT DEVICES OR SYSTEMS WITHOUT THE EXPRESS WRITTEN APPROVAL OF THE PRESIDENT OF NATIONAL SEMICONDUCTOR CORPORATION. As used herein:

1. Life support devices or systems are devices or systems which, (a) are intended for surgical implant into the body, or (b) support or sustain life, and whose failure to perform, when properly used in accordance with instructions for use provided in the labeling, can be reasonably expected to result in a significant injury to the user.

2. A critical component is any component of a life support device or system whose failure to perform can be reasonably expected to cause the failure of the life support device or system, or to affect its safety or effectiveness.

National Semiconductor Corporation	National Semiconductor GmbH	NS Japan Ltd.	National Semiconductor Hong Kong Ltd.	National Semicondutores Do Brasil Ltda.	National Semiconductor (Australia) PTY. Ltd.
2900 Semiconductor Drive	Westendstrasse 193-195	Sanseido Bldg. 5F	Southeast Asia Marketing	Av. Brig. Faria Lima, 830	21/3 High Street
P.O. Box 58090	D-8000 Munchen 21	4-15 Nishi Shinjuku	Austin Tower, 4th Floor	8 Andar	Bayswater, Victoria 3153
Santa Clara, CA 95052-8090	West Germany	Shinjuku-Ku,	22-26A Austin Avenue	01452 Sao Paulo, SP. Brasil	Australia
Tel: (408) 721-5000	Tel: (089) 5 70 95 01	Tokyo 160, Japan	Tsimshatsui, Kowloon, H.K.	Tel: (55/11) 212-5066	Tel: (03) 729-6333
TWX: (910) 339-9240	Telex: 522772	Tel: 3-299-7001	Tel: 3-7231290, 3-7243845	Telex: 391-1131931 NSBR BR	Telex: AA32096
		FAX: 3-299-7000	Cable: NSSEAMKTG		
			Telex: 52996 NSSEA HX		

National does not assume any responsibility for use of any circuitry described, no circuit patent licenses are implied and National reserves the right at any time without notice to change said circuitry and specifications.

Block Diagrams (Continued)

CD4053BM/CD4053BC

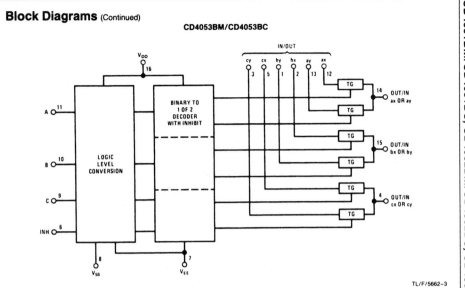

TL/F/5662-3

Truth Table

INPUT STATES				"ON" CHANNELS		
INHIBIT	C	B	A	CD4051B	CD4052B	CD4053B
0	0	0	0	0	0X, 0Y	cx, bx, ax
0	0	0	1	1	1X, 1Y	cx, bx, ay
0	0	1	0	2	2X, 2Y	cx, by, ax
0	0	1	1	3	3X, 3Y	cx, by, ay
0	1	0	0	4		cy, bx, ax
0	1	0	1	5		cy, bx, ay
0	1	1	0	6		cy, by, ax
0	1	1	1	7		cy, by, ay
1	*	*	*	NONE	NONE	NONE

*Don't Care condition.

National Semiconductor

MM54C90/MM74C90 4–Bit Decade Counter
MM54C93/MM74C93 4–Bit Binary Counter

General Description

The MM54C90/MM74C90 decade counter and the MM54C93/MM74C93 binary counter and complementary MOS (CMOS) integrated circuits constructed with N and P-channel enhancement mode transistors. The 4-bit decade counter can reset to zero or preset to nine by applying appropriate logic level on the R_{01}, R_{02}, R_{91} and R_{92} inputs. Also, a separate flip-flop on the A-bit enables the user to operate it as a divide–by–2, 5 or 10 frequency counter. The 4–bit binary counter can be reset to zero by applying high logic level on inputs R_{01} and R_{02}, and a separate flip-flop on the A-bit enables the user to operate it as a divide–by–2, –8, or –16 divider. Counting occurs on the negative going edge of the input pulse.

All inputs are protected against static discharge damage.

Features

- Wide supply voltage range   3V to 15V
- Guaranteed noise margin   1V
- High noise immunity   0.45 V_{CC} (typ.)
- Low power   fan out of 2
  TTL compatibility   driving 74L
- The MM54C93/MM74C93 follows the MM54L93/MM74L93 Pinout

Logic and Connection Diagrams

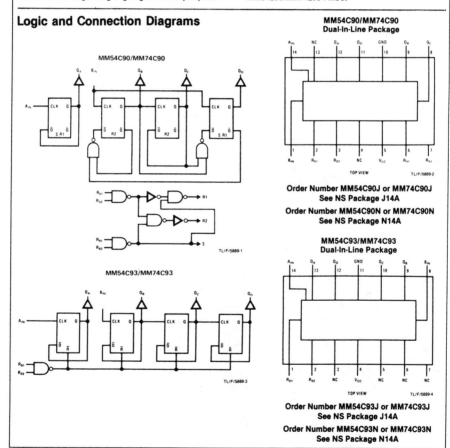

MM54C90/MM74C90

MM54C93/MM74C93

MM54C90/MM74C90
Dual-In-Line Package

TOP VIEW

Order Number MM54C90J or MM74C90J
See NS Package J14A

Order Number MM54C90N or MM74C90N
See NS Package N14A

MM54C93/MM74C93
Dual-In-Line Package

TOP VIEW

Order Number MM54C93J or MM74C93J
See NS Package J14A

Order Number MM54C93N or MM74C93N
See NS Package N14A

153

Switching Time Waveforms

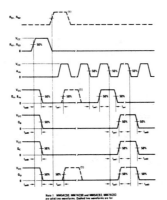

Note 1 MM54C90, MM74C90 and MM54C93, MM74C93 are valid line waveforms. Dashed line waveforms are for MM54C90/MM74C90 only.

TL/F/5889-7

Truth Tables

MM54C90/MM74C90 4-Bit Decade Counter

BCD Count Sequence

COUNT	OUTPUT			
	Q_D	Q_C	Q_B	Q_A
0	L	L	L	L
1	L	L	L	H
2	L	L	H	L
3	L	L	H	H
4	L	H	L	L
5	L	H	L	H
6	L	H	H	L
7	L	H	H	H
8	H	L	L	L
9	H	L	L	H

Output Q_A is connected to input B for BCD count.

H = High level
L = Low level
X = Irrelevant

Reset/Count Function Table

RESET INPUTS				OUTPUT			
R_{01}	R_{02}	R_{91}	R_{92}	Q_D	Q_C	Q_B	Q_A
H	H	L	X	L	L	L	L
H	H	X	L	L	L	L	L
X	X	H	H	H	L	L	H
X	L	X	L	COUNT			
L	X	L	X	COUNT			
L	X	X	L	COUNT			
X	L	L	X	COUNT			

MM54C93/MM74C93 4-Bit Binary Counter

Binary Count Sequence

COUNT	OUTPUT			
	Q_D	Q_C	Q_B	Q_A
0	L	L	L	L
1	L	L	L	H
2	L	L	H	L
3	L	L	H	H
4	L	H	L	L
5	L	H	L	H
6	L	H	H	L
7	L	H	H	H
8	H	L	L	L
9	H	L	L	H
10	H	L	H	L
11	H	L	H	H
12	H	H	L	L
13	H	H	L	H
14	H	H	H	L
15	H	H	H	H

Output Q_A is connected to input B for binary count sequence.

H = High level
L = Low level
X = Irrelevant

Reset/Count Function Table

RESET INPUTS		OUTPUT			
R_{01}	R_{02}	Q_D	Q_C	Q_B	Q_A
H	H	L	L	L	L
L	X	COUNT			
X	L	COUNT			

National Semiconductor Corporation
2900 Semiconductor Drive
P.O. Box 58090
Santa Clara, CA 95052-8090
Tel: (408) 721-5000
TWX: (910) 339-9240

National Semiconductor GmbH
Westendstrasse 193-195
D-8000 Munchen 21
West Germany
Tel: (089) 5 70 95 01
Telex: 522772

NS Japan Ltd.
Sanseido Bldg. 5F
4-15 Nishi Shinjuku
Shinjuku-Ku,
Tokyo 160, Japan
Tel: 3-299-7001
FAX: 3-299-7000

National Semiconductor Hong Kong Ltd.
Southeast Asia Marketing
Austin Tower, 4th Floor
22-26A Austin Avenue
Tsimshatsui, Kowloon, H.K.
Tel: 3-7231290, 3-7243645
Cable: NSSEAMKTG
Telex: 52996 NSSEA HX

National Semicondutores Do Brasil Ltda.
Av. Brig. Faria Lima, 830
8 Andar
01452 Sao Paulo, SP. Brasil
Tel: (55/11) 212-5066
Telex: 391-1131931 NSBR BR

National Semiconductor (Australia) PTY, Ltd.
21/3 High Street
Bayswater, Victoria 3153
Australia
Tel: (03) 729-6333
Telex: AA32096

National does not assume any responsibility for use of any circuitry described, no circuit patent licenses are implied and National reserves the right at any time without notice to change said circuitry and specifications.

154

 **National Semiconductor**

Industrial Blocks

LM555/LM555C Timer

General Description

The LM555 is a highly stable device for generating accurate time delays or oscillation. Additional terminals are provided for triggering or resetting if desired. In the time delay mode of operation, the time is precisely controlled by one external resistor and capacitor. For astable operation as an oscillator, the free running frequency and duty cycle are accurately controlled with two external resistors and one capacitor. The circuit may be triggered and reset on falling waveforms, and the output circuit can source or sink up to 200 mA or drive TTL circuits.

Features

- Direct replacement for SE555/NE555
- Timing from microseconds through hours
- Operates in both astable and monostable modes

- Adjustable duty cycle
- Output can source or sink 200 mA
- Output and supply TTL compatible
- Temperature stability better than 0.005% per °C
- Normally on and normally off output

Applications

- Precision timing
- Pulse generation
- Sequential timing
- Time delay generation
- Pulse width modulation
- Pulse position modulation
- Linear ramp generator

Schematic Diagram

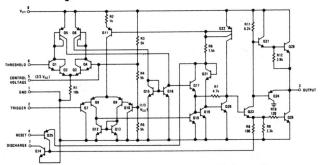

Connection Diagrams

Metal Can Package

TOP VIEW

Order Number LM555H, LM555CH
See NS Package H08C

Dual-In-Line Package

TOP VIEW

Order Number LM555CN
See NS Package N08B
Order Number LM555J or LM555CJ
See NS Package J08A

 National Semiconductor Corporation
2900 Semiconductor Drive
P.O. Box 58090
Santa Clara, CA 95052-8090
Tel: (408) 721-5000
TWX: (910) 339-9240

National Semiconductor GmbH
Westendstrasse 193-195
D-8000 Munchen 21
West Germany
Tel: (089) 5 70 95 01
Telex: 522772

NS Japan Ltd.
Sanseido Bldg. 5F
4-15 Nishi Shinjuku
Shinjuku-Ku,
Tokyo 160, Japan
Tel: 3-299-7001
FAX: 3-299-7000

National Semiconductor Hong Kong Ltd.
Southeast Asia Marketing
Austin Tower, 4th Floor
22-26A Austin Avenue
Tsimshatsui, Kowloon, H.K.
Tel: 3-7231290, 3-7243645
Cable: NSSEAMKTG
Telex: 52996 NSSEA HX

National Semicondutores Do Brasil Ltda.
Av. Brig. Faria Lima, 830
8 Andar
01452 Sao Paulo, SP. Brasil
Tel: (55/11) 212-5066
Telex: 391-1131931 NSBR BR

National Semiconductor (Australia) PTY. Ltd.
21/3 High Street
Bayswater, Victoria 3153
Australia
Tel: (03) 729-6333
Telex: AA32096

National does not assume any responsibility for use of any circuitry described, no circuit patent licenses are implied and National reserves the right at any time without notice to change said circuitry and specifications.

Applications Information

MONOSTABLE OPERATION

In this mode of operation, the timer functions as a one-shot (*Figure 1*). The external capacitor is initially held discharged by a transistor inside the timer. Upon application of a negative trigger pulse of less than 1/3 V_{CC} to pin 2, the flip-flop is set which both releases the short circuit across the capacitor and drives the output high.

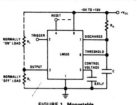

FIGURE 1. Monostable

The voltage across the capacitor then increases exponentially for a period of t = 1.1 R_AC, at the end of which time the voltage equals 2/3 V_{CC}. The comparator then resets the flip-flop which in turn discharges the capacitor and drives the output to its low state. *Figure 2* shows the waveforms generated in this mode of operation. Since the charge and the threshold level of the comparator are both directly proportional to supply voltage, the timing internal is independent of supply.

V_{CC} = 5V
TIME = 0.1 ms/DIV.
R_A = 9.1kΩ
C = 0.01μF

Top Trace: Input 5V/Div.
Middle Trace: Output 5V/Div.
Bottom Trace: Capacitor Voltage 2V/Div.

FIGURE 2. Monostable Waveforms

During the timing cycle when the output is high, the further application of a trigger pulse will not effect the circuit. However the circuit can be reset during this time by the application of a negative pulse to the reset terminal (pin 4). The output will then remain in the low state until a trigger pulse is again applied.

When the reset function is not in use, it is recommended that it be connected to V_{CC} to avoid any possibility of false triggering.

Figure 3 is a nomograph for easy determination of R, C values for various time delays.

NOTE: In monostable operation, the trigger should be driven high before the end of timing cycle.

ASTABLE OPERATION

If the circuit is connected as shown in *Figure 4* (pins 2 and 6 connected) it will trigger itself and free run as a

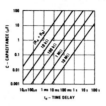

FIGURE 3. Time Delay

multivibrator. The external capacitor charges through R_A + R_B and discharges through R_B. Thus the duty cycle may be precisely set by the ratio of these two resistors.

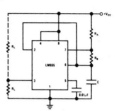

FIGURE 4. Astable

In this mode of operation, the capacitor charges and discharges between 1/3 V_{CC} and 2/3 V_{CC}. As in the triggered mode, the charge and discharge times, and therefore the frequency are independent of the supply voltage.

Figure 5 shows the waveforms generated in this mode of operation.

V_{CC} = 5V
TIME = 20μs/DIV.
R_A = 3.9 kΩ
R_B = 3 kΩ
C = 0.01μF

Top Trace: Output 5V/Div.
Bottom Trace: Capacitor Voltage 1V/Div.

FIGURE 5. Astable Waveforms

The charge time (output high) is given by:
$$t_1 = 0.693 \ (R_A + R_B) \ C$$

And the discharge time (output low) by:
$$t_2 = 0.693 \ (R_B) \ C$$

Thus the total period is:
$$T = t_1 + t_2 = 0.693 \ (R_A + 2R_B) \ C$$

LIFE SUPPORT POLICY

 National Semiconductor Corporation
2900 Semiconductor Drive
P.O. Box 58090
Santa Clara, CA 95052-8090
Tel: (408) 721-5000
TWX: (910) 339-9240

National Semiconductor GmbH
Westendstrasse 193-195
D-8000 Munchen 21
West Germany
Tel: (089) 5 70 95 01
Telex: 522772

NS Japan Ltd.
Sanseido Bldg. 5F
4-15 Nishi Shinjuku
Shinjuku-Ku,
Tokyo 160, Japan
Tel: 3-299-7001
FAX: 3-299-7000

National Semiconductor Hong Kong Ltd.
Southeast Asia Marketing
Austin Tower, 4th Floor
22-26A Austin Avenue
Tsimshatsui, Kowloon, H.K.
Tel: 3-7231290, 3-7243645
Cable: NSSEAMKTG.
Telex: 52996 NSSEA HX

National Semicondutores Do Brasil Ltda.
Av. Brig. Faria Lima, 830
8 Andar
01452 Sao Paulo, SP. Brasil
Tel: (55/11) 212-5066
Telex: 391-1131931 NSBR BR

National Semiconductor (Australia) PTY. Ltd.
21/3 High Street
Bayswater, Victoria 3153
Australia
Tel: (03) 729-6333
Telex: AA32096

Applications Information (Continued)

The frequency of oscillation is:

$$f = \frac{1}{T} = \frac{1.44}{(R_A + 2R_B)\,C}$$

Figure 6 may be used for quick determination of these RC values.

The duty cycle is:

$$D = \frac{R_B}{R_A + 2R_B}$$

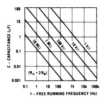

FIGURE 6. Free Running Frequency

FREQUENCY DIVIDER

The monostable circuit of *Figure 1* can be used as a frequency divider by adjusting the length of the timing cycle. *Figure 7* shows the waveforms generated in a divide by three circuit.

V_CC = 5V
TIME = 20μs/DIV.
R_A = 9.1 kΩ
C = 0.01μF

Top Trace: Input 4V/Div.
Middle Trace: Output 2V/Div.
Bottom Trace: Capacitor 2V/Div.

FIGURE 7. Frequency Divider

PULSE WIDTH MODULATOR

When the timer is connected in the monostable mode and triggered with a continuous pulse train, the output pulse width can be modulated by a signal applied to pin 5. *Figure 8* shows the circuit, and in *Figure 9* are some waveform examples.

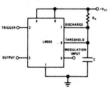

FIGURE 8. Pulse Width Modulator

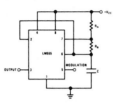

V_CC = 5V
TIME = 0.2 ms/DIV.
R_A = 9.1 kΩ
C = 0.01μF

Top Trace: Modulation 1V/Div.
Bottom Trace: Output 2V/Div.

FIGURE 9. Pulse Width Modulator

PULSE POSITION MODULATOR

This application uses the timer connected for astable operation, as in *Figure 10*, with a modulating signal again applied to the control voltage terminal. The pulse position varies with the modulating signal, since the threshold voltage and hence the time delay is varied. *Figure 11* shows the waveforms generated for a triangle wave modulation signal.

FIGURE 10. Pulse Position Modulator

V_CC = 5V
TIME = 0.1 ms/DIV.
R_A = 3.9 kΩ
R_B = 3 kΩ
C = 0.01μF

Top Trace: Modulation Input 1V/Div.
Bottom Trace: Output 2V/Div.

FIGURE 11. Pulse Position Modulator

LINEAR RAMP

When the pullup resistor, R_A, in the monostable circuit is replaced by a constant current source, a linear ramp is

National Semiconductor Corporation	National Semiconductor GmbH	NS Japan Ltd.	National Semiconductor Hong Kong Ltd.	National Semicondutores Do Brasil Ltda.	National Semiconductor (Australia) PTY. Ltd.
2900 Semiconductor Drive	Westendstrasse 193-195	Sanseido Bldg. 5F	Southeast Asia Marketing	Av. Brig. Faria Lima, 830	21/3 High Street
P.O. Box 58090	D-8000 Munchen 21	4-15 Nishi Shinjuku	Austin Tower, 4th Floor	8 Andar	Bayswater, Victoria 3153
Santa Clara, CA 95052-8090	West Germany	Shinjuku-Ku.	22-26A Austin Avenue	01452 Sao Paulo, SP. Brasil	Australia
Tel: (408) 721-5000	Tel: (089) 5 70 95 01	Tokyo 160, Japan	Tsimshatsui, Kowloon, H.K.	Tel: (55/11) 212-5066	Tel: (03) 729-6333
TWX: (910) 339-9240	Telex: 522772	Tel: 3-299-7001	Tel: 3-7231290, 3-7243645	Telex: 391-1131931 NSBR BR	Telex: AA32096
		FAX: 3-299-7000	Cable: NSSEAMKTG		
			Telex: 52996 NSSEA HX		

National does not assume any responsibility for use of any circuitry described, no circuit patent licenses are implied and National reserves the right at any time without notice to change said circuitry and specifications.

Applications Information (Continued)

generated. *Figure 12* shows a circuit configuration that will perform this function.

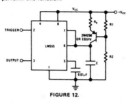

FIGURE 12.

Figure 13 shows waveforms generated by the linear ramp.

The time interval is given by:

$$T = \frac{2/3\,V_{CC}\,R_E\,(R_1 + R_2)\,C}{R_1\,V_{CC} - V_{BE}\,(R_1 + R_2)}$$

$$V_{BE} \simeq 0.6V$$

V_{CC} = 5V
TIME = 20μs/DIV
R_1 = 47 kΩ
R_2 = 100 kΩ
R_E = 2.7 kΩ
C = 0.01μF

Top Trace: Input 2V/Div.
Middle Trace: Output 5V/Div.
Bottom Trace: Capacitor Voltage 1V/Div.

FIGURE 13. Linear Ramp

50% DUTY CYCLE OSCILLATOR

For a 50% duty cycle, the resistors R_A and R_B may be connected as in *Figure 14*. The time period for the out-

put high is the same as previous, $t_1 = 0.693\,R_A\,C$. For the output low it is $t_2 =$

$$[(R_A\,R_B)/(R_A + R_B)]\,C\ln\left[\frac{R_B - 2R_A}{2R_B - R_A}\right]$$

Thus the frequency of oscillation is $f = \dfrac{1}{t_1 + t_2}$

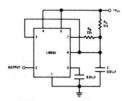

FIGURE 14. 50% Duty Cycle Oscillator

Note that this circuit will not oscillate if R_B is greater than 1/2 R_A because the junction of R_A and R_B cannot bring pin 2 down to 1/3 V_{CC} and trigger the lower comparator.

ADDITIONAL INFORMATION

Adequate power supply bypassing is necessary to protect associated circuitry. Minimum recommended is 0.1μF in parallel with 1μF electrolytic.

Lower comparator storage time can be as long as 10μs when pin 2 is driven fully to ground for triggering. This limits the monostable pulse width to 10μs minimum.

Delay time reset to output is 0.47μs typical. Minimum reset pulse width must be 0.3μs, typical.

Pin 7 current switches within 30 ns of the output (pin 3) voltage.

MC1648/MC1648M

VOLTAGE-CONTROLLED OSCILLATOR

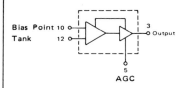

Bias Point 10

Tank 12

3 Output

5 AGC

Input Capacitance = 6 pF typ
Maximum Series Resistance for L (External
  Inductance) = 50 Ω typ
Power Dissipation = 150 mW typ/pkg
  (+5.0 Vdc Supply)
Maximum Output Frequency = 225 MHz typ

The MC1648 requires an external parallel tank circuit consisting of the inductor (L) and capacitor (C).

A varactor diode may be incorporated into the tank circuit to provide a voltage variable input for the oscillator (VCO). The MC1648 was designed for use in the Motorola Phase-Locked Loop shown in Figure 9. This device may also be used in many other applications requiring a fixed or variable frequency clock source of high spectral purity. (See Figure 2.)

The MC1648 may be operated from a +5.0 Vdc supply or a –5.2 Vdc supply, depending upon system requirements.

Supply Voltage	Gnd Pins	Supply Pins
+5.0 Vdc	7, 8	1, 14
–5.2 Vdc	1, 14	7, 8

L SUFFIX
CERAMIC PACKAGE
CASE 632

P SUFFIX
PLASTIC PACKAGE
CASE 646

F SUFFIX
CERAMIC PACKAGE
CASE 607

FIGURE 1 – CIRCUIT SCHEMATIC

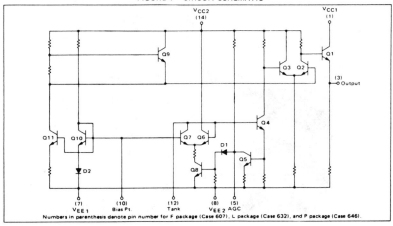

Numbers in parenthesis denote pin number for F package (Case 607), L package (Case 632), and P package (Case 646).

Copyright of Motorola, Inc. Used by Permission.

159

MC1648/MC1648M

TEST VOLTAGE/CURRENT VALUES

@ Test Temperature	VIHmax (Volts)	VILmin (Volts)	VCC (Volts)	IL (mAdc)
MC1648				
−30°C	+2.00	+1.50	5.0	−5.0
+25°C	+1.85	+1.35	5.0	−5.0
+85°C	+1.70	+1.20	5.0	−5.0
MC1648M				
−55°C	+2.07	+1.57	5.0	−5.0
+25°C	+1.85	+1.35	5.0	−5.0
+125°C	+1.60	+1.10	5.0	−5.0

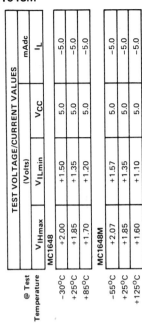

FIGURE 2 – SPECTRAL PURITY OF SIGNAL AT OUTPUT

+5.0 Vdc

L: Micro Metal torroid #T20-22, 8 turns #30 Enamled Copper wire.

C = 3.0 – 35 pF

10 µF
0.1 µF
1200*
Signal Under Test
0.1 µF

*The 1200 ohm resistor and the scope termination impedance constitute a 25:1 attenuator probe. Coax shall be CT-070-50 or equivalent.

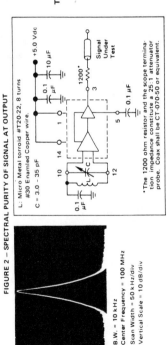

B.W. = 10 kHz
Center Frequency = 100 MHz
Scan Width = 50 kHz/div
Vertical Scale = 10 dB/div

ELECTRICAL CHARACTERISTICS
Supply Voltage = +5.0 Volts

Characteristic	Symbol	−55°C Min	−55°C Max	−30°C Min	−30°C Max	+25°C Min	+25°C Typ	+25°C Max	+85°C Min	+85°C Typ	+85°C Max	+125°C Min	+125°C Typ	+125°C Max	Unit	Conditions
Power Supply Drain Current	I_E	–	–	–	–	–	–	41	–	–	–	–	–	–	mAdc	Inputs and outputs open.
Logic "1" Output Voltage	V_{OH}	3.92	4.13	3.955	4.185	4.04	–	4.25	4.11	–	4.36	4.16	–	4.40	Vdc	$V_{IL}min$ to Pin 12, I_L @ Pin 3.
Logic "0" Output Voltage	V_{OL}	3.13	3.38	3.16	3.40	3.20	–	3.43	3.22	–	3.475	3.23	–	3.51	Vdc	$V_{IH}max$ to Pin 12, I_L @ Pin 3.
Bias Voltage	V_{Bias}*	1.67	1.97	1.60	1.90	1.45	–	1.75	1.30	–	1.60	1.20	–	1.50	Vdc	$V_{IL}min$ to Pin 12.
Peak-to-Peak Tank Voltage	V_{P-P}	–	–	–	–	–	400	–	–	–	–	–	–	–	mV	See Figure 3.
Output Duty Cycle	V_{DC}	–	–	–	–	–	50	–	–	–	–	–	–	–	%	See Figure 3.
Oscillation Frequency	f_{max}**	–	225	–	225	200	225	–	–	225	–	–	225	–	MHz	See Figure 3.

*This measurement guarantees the dc potential at the bias point for purposes of incorporating a varactor tuning diode at this point.

**Frequency variation over temperature is a direct function of the ΔC/Δ Temperature and ΔL/Δ Temperature.

TRANSFER CHARACTERISTICS IN THE VOLTAGE CONTROLLED MODE
USING EXTERNAL VARACTOR DIODE AND COIL. $T_A = 25^\circ C$
FIGURE 6

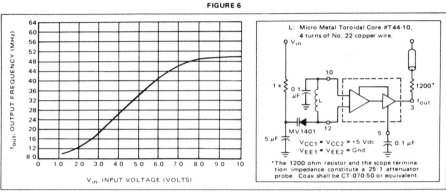

FIGURE 7

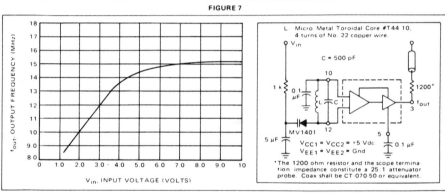

FIGURE 8

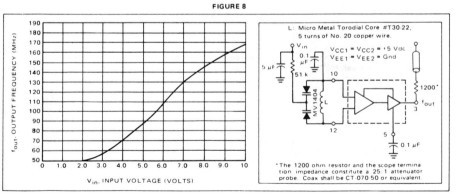

Typical transfer characteristics for the oscillator in the voltage controlled mode are shown in Figures 6, 7, and 8. Figures 6 and 8 show transfer characteristics employing only the capacitance of the varactor diode (plus the input capacitance of the oscillator, 6 pF typical). Figure 7 illustrates the oscillator operating in a voltage controlled mode with the output frequency range limited. This is achieved by adding a capacitor in parallel with the tank circuit as shown. The 1 kΩ resistor in Figures 6 and 7 is used to protect the varactor diode during testing. It is not necessary as long as the dc input voltage does not cause the diode to become forward biased. The larger-valued resistor (51 kΩ) in Figure 8 is required to provide isolation for the high-impedance junctions of the two varactor diodes.

The tuning range of the oscillator in the voltage controlled mode may be calculated as:

$$\frac{f_{max}}{f_{min}} = \frac{\sqrt{C_D(max) + C_S}}{\sqrt{C_D(min) + C_S}}$$

$$\text{where } f_{min} = \frac{1}{2\pi\sqrt{L(C_D(max) + C_S)}}$$

C_S = shunt capacitance (input plus external capacitance).

C_D = varactor capacitance as a function of bias voltage.

Good RF and low-frequency bypassing is necessary on the power supply pins. (See Figure 2.)

Capacitors (C1 and C2 of Figure 4) should be used to bypass the AGC point and the VCO input (varactor diode), guaranteeing only dc levels at these points.

For output frequency operation between 1 MHz and 50 MHz a 0.1 μF capacitor is sufficient for C1 and C2. At higher frequencies, smaller values of capacitance should be used; at lower frequencies, larger values of capacitance. At high frequencies the value of bypass capacitors depends directly upon the physical layout of the system. All bypassing should be as close to the package pins as possible to minimize unwanted lead inductance.

The peak-to-peak swing of the tank circuit is set internally by the AGC circuitry. Since voltage swing of the tank circuit provides the drive for the output buffer, the AGC potential directly affects the output waveform. If it is desired to have a sine wave at the output of the MC1648, a series resistor is tied from the AGC point to the most negative power potential (ground if +5.0 volt supply is used, −5.2 volts if a negative supply is used) as shown in Figure 10.

At frequencies above 100 MHz typ, it may be desirable to increase the tank circuit peak-to-peak voltage in order to shape the signal at the output of the MC1648. This is accomplished by tying a series resistor (1 kΩ minimum) from the AGC to the most positive power potential (+5.0 volts if a +5.0 volt supply is used, ground if a −5.2 volt supply is used). Figure 11 illustrates this principle.

APPLICATIONS INFORMATION

The phase locked loop shown in Figure 9 illustrates the use of the MC1648 as a voltage controlled oscillator. The figure illustrates a frequency synthesizer useful in tuners for FM broadcast, general aviation, maritime and landmobile communications, amateur and CB receivers. The system operates from a single +5.0 Vdc supply, and requires no internal translations, since all components are compatible.

Frequency generation of this type offers the advantages of single crystal operation, simple channel selection, and elimination of special circuitry to prevent harmonic lockup. Additional features include dc digital switching

(preferable over RF switching with a multiple crystal system), and a broad range of tuning (up to 150 MHz, the range being set by the varactor diode).

The output frequency of the synthesizer loop is determined by the reference frequency and the number programmed at the programmable counter; $f_{out} = Nf_{ref}$. The channel spacing is equal to frequency (f_{ref}).

For additional information on applications and designs for phase locked-loops and digital frequency synthesizers, see Motorola Application Notes AN-532A, AN-535, AN-553, AN-564 or AN594.

Signetics

TDA7000
Single-Chip FM Radio Circuit

Product Specification

Linear Products

DESCRIPTION

The TDA7000 is a monolithic integrated circuit for mono FM portable radios where a minimum of peripheral components is important (small dimensions and low costs).

The IC has an FLL (Frequency-Locked Loop) system with an intermediate frequency of 70kHz. The IF selectivity is obtained by active RC filters. The only function which needs tuning is the resonant circuit for the oscillator which selects the reception frequency. Spurious reception is avoided by means of a mute circuit, which also eliminates weak, noisy input signals. Special precautions are taken to meet the radiation requirements.

FEATURES

- RF input stage
- Mixer
- Local oscillator
- IF amplifier/limiter
- Phase demodulator
- Mute detector
- Mute switch

APPLICATIONS

- Mono FM Portable Radios
- LAN
- Data Receivers
- SCA Receiver

PIN CONFIGURATION

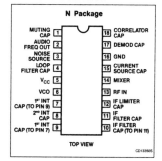

ORDERING INFORMATION

DESCRIPTION	TEMPERATURE RANGE	ORDER CODE
18-Pin Plastic DIP (SOT-102HE)	0 to +70°C	TDA7000N

ABSOLUTE MAXIMUM RATINGS

SYMBOL	PARAMETER	RATING	UNIT
V_{CC}	Supply voltage (Pin 5)	12	V
V_{6-5}	Oscillator voltage (Pin 6)	$V_{CC} - 0.5$ to $V_{CC} + 0.5$	V
P_{TOT}	Total power dissipation	See derating curve Figure 1	
T_{STG}	Storage temperature range	-55 to +150	°C
T_A	Operating ambient temperature range	0 to +60	°C

163

DC ELECTRICAL CHARACTERISTICS V_{CC} = 4.5V; T_A = 25°C; measured in Figure 3, unless otherwise specified.

SYMBOL	PARAMETER	TEST CONDITIONS	LIMITS			UNIT
			Min	Typ	Max	
V_{CC}	Supply voltage	(Pin 5)	2.7	4.5	10	V
I_{CC}	Supply current	V_{CC} = 4.5V		8		mA
I_6	Oscillator current	(Pin 6)		280		μA
V_{14-16}	Voltage	(Pin 14)		1.35		V
I_2	Output current	(Pin 2)		60		μA
V_{2-16}	Output voltage	(Pin 2) R_L = 22kΩ		1.3		V

AC ELECTRICAL CHARACTERISTICS V_{CC} = 4.5V; T_A = 25°C; measured in Figure 3 (mute switch open, enabled); f_{RF} = 96MHz (tuned to max. signal at 5μV EMF) modulated with Δf = ± 22.5kHz; f_M = 1kHz; EMF = 0.2mV (EMF voltage at a source impedance of 75Ω); RMS noise voltage measured unweighted (f = 300Hz to 20kHz), unless otherwise specified.

SYMBOL	PARAMETER	TEST CONDITIONS	LIMITS			UNIT
			Min	Typ	Max	
EMF	Sensitivity (see Figure 2) (EMF voltage)	−3dB limiting; muting disabled		1.5		μV
		−3dB muting		6		
		S/N = 26dB		5.5		
EMF	Signal handling (EMF voltage)	THD < 10%; Δf = ± 75kHz		200		mV
S/N	Signal-to-noise ratio			60		dB
THD	Total harmonic distortion	Δf = ± 22.5kHz		0.7		%
		Δf = ± 75kHz		2.3		
AMS	AM suppression of output voltage	(ratio of the AM output signal referred to the FM output signal) FM signal: f_M = 1kHz; Δf = ± 75kHz AM signal: f_M = 1kHz; m = 80%		50		dB
RR	Ripple rejection	(ΔV_{CC} = 100mV; f = 1kHz)		10		dB
$V_{6-5(RMS)}$	Oscillator voltage (RMS value)	(Pin 6)		250		mV
Δf_{OSC}	Variation of oscillator frequency	Supply voltage (ΔV_{CC} = 1V)		60		kHz/V
S_{+300}	Selectivity			45		dB
S_{-300}				35		
Δf_{RF}	AFC range			± 300		kHz
BW	Audio bandwidth	ΔV_O = 3dB measured with pre-emphasis (t = 50μs)		10		kHz
$V_{O\ RMS}$	AF output voltage (RMS value)	R_L = 22kΩ		75		mV
R_L	Load resistance	V_{CC} = 4.5V			22	kΩ
		V_{CC} = 9.0V			47	

NOTES:
1. The muting system can be disabled by feeding a current of about 20μA into Pin 1.
2. The interstation noise level can be decreased by choosing a low-value capacitor at Pin 3. Silent tuning can be achieved by omitting this capacitor.

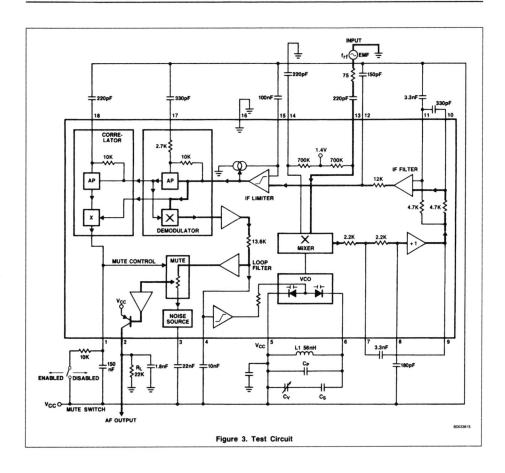

Figure 3. Test Circuit

 **National Semiconductor**

LM386 Low Voltage Audio Power Amplifier

General Description

The LM386 is a power amplifier designed for use in low voltage consumer applications. The gain is internally set to 20 to keep external part count low, but the addition of an external resistor and capacitor between pins 1 and 8 will increase the gain to any value up to 200.

The inputs are ground referenced while the output is automatically biased to one half the supply voltage. The quiescent power drain is only 24 milliwatts when operating from a 6 volt supply, making the LM386 ideal for battery operation.

Features

- Battery operation
- Minimum external parts
- Wide supply voltage range   4V−12V or 5V−18V
- Low quiescent current drain   4 mA

- Voltage gains from 20 to 200
- Ground referenced input
- Self-centering output quiescent voltage
- Low distortion
- Eight pin dual-in-line package

Applications

- AM-FM radio amplifiers
- Portable tape player amplifiers
- Intercoms
- TV sound systems
- Line drivers
- Ultrasonic drivers
- Small servo drivers
- Power converters

Equivalent Schematic and Connection Diagrams

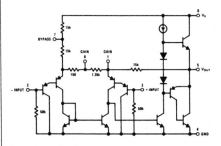

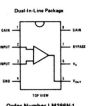

Dual-In-Line Package

TOP VIEW

Order Number LM386N-1, LM386N-3 or LM386N-4
See NS Package N08B

Typical Applications

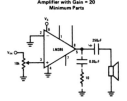

Amplifier with Gain = 20
Minimum Parts

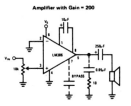

Amplifier with Gain = 200

LIFE SUPPORT POLICY

NATIONAL'S PRODUCTS ARE NOT AUTHORIZED FOR USE AS CRITICAL COMPONENTS IN LIFE SUPPORT DEVICES OR SYSTEMS WITHOUT THE EXPRESS WRITTEN APPROVAL OF THE PRESIDENT OF NATIONAL SEMICONDUCTOR CORPORATION. As used herein:

1. Life support devices or systems are devices or systems which, (a) are intended for surgical implant into the body, or (b) support or sustain life, and whose failure to perform, when properly used in accordance with instructions for use provided in the labeling, can be reasonably expected to result in a significant injury to the user.

2. A critical component is any component of a life support device or system whose failure to perform can be reasonably expected to cause the failure of the life support device or system, or to affect its safety or effectiveness.

 National Semiconductor Corporation
2900 Semiconductor Drive
P.O. Box 58090
Santa Clara, CA 95052-8090
Tel: (408) 721-5000
TWX: (910) 339-9240

National Semiconductor GmbH
Westendstrasse 193-195
D-8000 Munchen 21
West Germany
Tel: (089) 5 70 95 01
Telex: 522772

NS Japan Ltd.
Sanseido Bldg. 5F
4-15 Nishi Shinjuku
Shinjuku-Ku,
Tokyo 160, Japan
Tel: 3-299-7001
FAX: 3-299-7000

National Semiconductor Hong Kong Ltd.
Southeast Asia Marketing
Austin Tower, 4th Floor
22-26A Austin Avenue
Tsimshatsui, Kowloon, H.K.
Tel: 3-7231290, 3-7243645
Cable: NSSEAMKTG
Telex: 52996 NSSEA HX

National Semicondutores Do Brasil Ltda.
Av. Brig. Faria Lima, 830
8 Andar
01452 Sao Paulo, SP. Brasil
Tel: (55/11) 212-5066
Telex: 391-1131931 NSBR BR

National Semiconductor (Australia) PTY. Ltd.
21/3 High Street
Bayswater, Victoria 3153
Australia
Tel: (03) 729-6333
Telex: AA32096

National does not assume any responsibility for use of any circuitry described, no circuit patent licenses are implied and National reserves the right at any time without notice to change said circuitry and specifications.

Signetics

NE/SE565
Phase-Locked Loop

Product Specification

Linear Products

DESCRIPTION

The NE/SE565 Phase-Locked Loop (PLL) is a self-contained, adaptable filter and demodulator for the frequency range from 0.001Hz to 500kHz. The circuit comprises a voltage-controlled oscillator of exceptional stability and linearity, a phase comparator, an amplifier and a low pass filter as shown in the Block Diagram. The center frequency of the PLL is determined by the free-running frequency of the VCO; this frequency can be adjusted externally with a resistor or a capacitor. The low pass filter, which determines the capture characteristics of the loop, is formed by an internal resistor and an external capacitor.

FEATURES

- Highly stable center frequency (200ppm/°C typ.)
- Wide operating voltage range (± 6V to ± 12V)
- Highly linear demodulated output (0.2% typ.)
- Center frequency programming by means of a resistor or capacitor, voltage or current
- TTL and DTL compatible square wave output; loop can be opened to insert digital frequency divider
- Highly linear triangle wave output
- Reference output for connection of comparator in frequency discriminator
- Bandwidth adjustable from < ± 1% to > ± 60%
- Frequency adjustable over 10 to 1 range with same capacitor

PIN CONFIGURATIONS

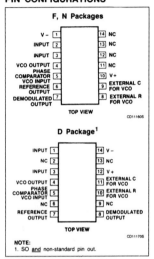

APPLICATIONS

- Frequency shift keying
- Modems
- Telemetry receivers
- Tone decoders
- SCA receivers
- Wide-band FM discriminators
- Data synchronizers
- Tracking filters
- Signal restoration
- Frequency multiplication & division

BLOCK DIAGRAM

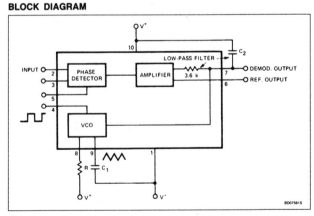

DC AND AC ELECTRICAL CHARACTERISTICS T_A = 25°C, V_{CC} = ± 6V, unless otherwise specified.

SYMBOL	PARAMETER	TEST CONDITIONS	SE565			NE565			UNIT
			Min	Typ	Max	Min	Typ	Max	
Supply requirements									
V_{CC}	Supply voltage		± 6		± 12	± 6		± 12	V
I_{CC}	Supply current			8	12.5		8	12.5	mA
Input characteristics									
	Input impedance[1]		7	10		5	10		kΩ
	Input level required for tracking	f_O = 50kHz, ± 10% frequency deviation	10			10			mV_{RMS}
VCO characteristics									
f_C	Center frequency Maximum value distribution[2]	Distribution taken about f_O = 50kHz, R_1 = 5.0kΩ, C_1 = 1200pF	300	500			500		kHz
			−10	0	+10	−30	0	+30	%
	Drift with temperature	f_O = 50kHz		500			600		ppm/°C
	Drift with supply voltage	f_O = 50kHz, V_{CC} = ± 6 to ± 7V		0.1	1.0		0.2	1.5	%/V
	Triangle wave output voltage level linearity		1.9	2.4 0.2	3	1.9	2.4 0.5	3	V_{P-P} %
	Square wave logical ''1'' output voltage logical ''0'' output voltage	f_O = 50kHz f_O = 50kHz	+ 4.9	+ 5.2 −0.2	+ 0.2	+ 4.9	+ 5.2 −0.2	+ 0.2	V V
	Duty cycle	f_O = 50kHz	45	50	55	40	50	60	%
t_R	Rise time			20	100		20		ns
t_F	Fall time			50	200		50		ns
I_{SINK}	Output current (sink)		0.6	1		0.6	1		mA
I_{SOURCE}	Output current (source)		5	10		5	10		mA
Demodulated output characteristics									
V_{OUT}	Output voltage level	Measured at Pin 7	4.25	4.5	4.75	4.0	4.5	5.0	V
	Maximum voltage swing[3]			2			2		V_{P-P}
	Output voltage swing	± 10% frequency deviation	250	300		200	300		mV_{P-P}
THD	Total harmonic distortion			0.2	0.75		0.4	1.5	%
	Output impedance[4]			3.6			3.6		kΩ
V_{OS}	Offset voltage (V6 – V7)			30	100		50	200	mV
	Offset voltage vs temperature (drift)			50			100		μV/°C
	AM rejection		30	40			40		dB

NOTES:
1. Both input terminals (Pins 2 and 3) must receive identical DC bias. This bias may range from 0V to −4V.
2. The external resistance for frequency adjustment (R_1) must have a value between 2kΩ and 20kΩ.
3. Output voltage swings negative as input frequency increases.
4. Output not buffered.

TYPICAL PERFORMANCE CHARACTERISTICS

Power Supply Current
as a Function of
Supply Voltage

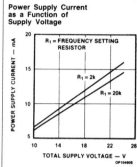

VCO Conversion Gain

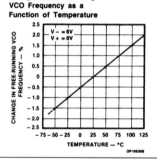

Lock Range
as a Function of
Input Voltage

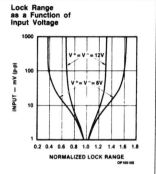

Lock Range
as a Function of
Gain Setting Resistance
(Pins 6 – 7)

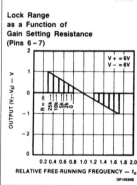

Change in Free-Running
VCO Frequency as a
Function of Temperature

VCO Output
Waveform

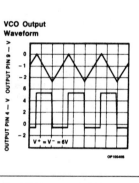

DESIGN FORMULAS
(See Figure 1)

Free-running frequency of VCO:

$$f_O \simeq \frac{1.2}{4R_1C_1} \text{ in Hz}$$

Lock range: $f_L = \pm \dfrac{8f_O}{V_{CC}}$ in Hz

Capture range: $f_C \simeq \pm \dfrac{1}{2\pi} \sqrt{\dfrac{2\pi f_L}{\tau}}$

where $\tau = (3.6 \times 10^3) \times C_2$

TYPICAL APPLICATIONS
FM Demodulation

The 565 Phase-Locked Loop is a general purpose circuit designed for highly linear FM demodulation. During lock, the average DC level of the phase comparator output signal is directly proportional to the frequency of the input signal. As the input frequency shifts, it is this output signal which causes the VCO to shift its frequency to match that of the input. Consequently, the linearity of the phase comparator output with frequency is determined by the voltage-to-frequency transfer function of the VCO.

Because of its unique and highly linear VCO, the 565 PLL can lock to and track an input signal over a very wide bandwidth (typically ±60%) with very high linearity (typically within 0.5%).

A typical connection diagram is shown in Figure 1. The VCO free-running frequency is given approximately by

$$f_O \simeq \frac{1.2}{4R_1C_1}$$

and should be adjusted to be at the center of the input signal frequency range. C_1 can be any value, but R_1 should be within the range of 2000 to 20,000Ω with an optimum value on the order of 4000Ω. The source can be direct coupled if the DC resistances seen from Pins 2 and 3 are equal and there is no DC voltage difference between the pins. A short between

Pins 4 and 5 connects the VCO to the phase comparator. Pin 6 provides a DC reference voltage that is close to the DC potential of the demodulated output (Pin 7). Thus, if a resistance is connected between Pins 6 and 7, the gain of the output stage can be reduced with little change in the DC voltage level at the output. This allows the lock range to be decreased with little change in the free-running frequency. In this manner the lock range can be decreased from ±60% of f_O to approximately ±20% of f_O (at ±6V).

A small capacitor (typically 0.001µF) should be connected between Pins 7 and 8 to eliminate possible oscillation in the control current source.

A single-pole loop filter is formed by the capacitor C2, connected between Pin 7 and the positive supply, and an internal resistance of approximately 3600Ω.

169

Signetics

LM193/A/293/A/393/A/2903
Low Power Dual Voltage Comparator

Linear Products

Product Specification

DESCRIPTION

The LM193 series consists of two independent precision voltage comparators with an offset voltage specification as low as 2.0mV max. for two comparators which were designed specifically to operate from a single power supply over a wide range of voltages. Operation from split power supplies is also possible and the low power supply current drain is independent of the magnitude of the power supply voltage. These comparators also have a unique characteristic in that the input common-mode voltage range includes ground, even though operated from a single power supply voltage.

The LM193 series was designed to directly interface with TTL and CMOS. When operated from both plus and minus power supplies, the LM193 series will directly interface with MOS logic where their low power drain is a distinct advantage over standard comparators.

FEATURES

- **Wide single supply voltage range $2.0V_{DC}$ to $36V_{DC}$ or dual supplies $\pm 1.0V_{DC}$, to $\pm 18V_{DC}$**
- **Very low supply current drain (0.8mA) independent of supply voltage (2.0mW/comparator at $5.0V_{DC}$)**
- **Low input biasing current 25nA**
- **Low input offset current $\pm 5nA$ and offset voltage $\pm 2mV$**
- **Input common-mode voltage range includes ground**
- **Differential input voltage range equal to the power supply voltage**
- **Low output 250mV at 4mA saturation voltage**
- **Output voltage compatible with TTL, DTL, ECL, MOS and CMOS logic systems**

APPLICATIONS

- **A/D converters**
- **Wide range VCO**
- **MOS clock generator**
- **High voltage logic gate**
- **Multivibrators**

PIN CONFIGURATION

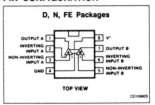

D, N, FE Packages

TOP VIEW

EQUIVALENT CIRCUIT

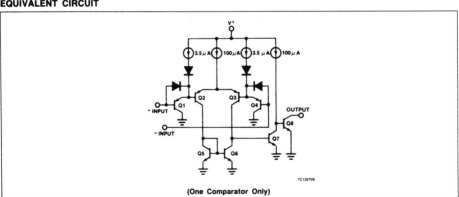

(One Comparator Only)

Signetics

NE/SE567
Tone Decoder/Phase-Locked Loop

Product Specification

DESCRIPTION

The NE/SE567 tone and frequency decoder is a highly stable phase-locked loop with synchronous AM lock detection and power output circuitry. Its primary function is to drive a load whenever a sustained frequency within its detection band is present at the self-biased input. The bandwidth center frequency and output delay are independently determined by means of four external components.

FEATURES

- Wide frequency range (.01Hz to 500kHz)
- High stability of center frequency
- Independently controllable bandwidth (up to 14%)
- High out-band signal and noise rejection
- Logic-compatible output with 100mA current sinking capability
- Inherent immunity to false signals

- Frequency adjustment over a 20–to-1 range with an external resistor
- Military processing available

APPLICATIONS

- Touch-Tone® decoding
- Carrier current remote controls
- Ultrasonic controls (remote TV, etc.)
- Communications paging
- Frequency monitoring and control
- Wireless intercom
- Precision oscillator

PIN CONFIGURATIONS

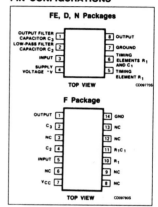

BLOCK DIAGRAM

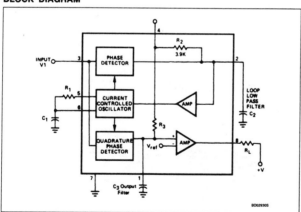

®Touch-Tone is a registered trademark of AT & T.

TYPICAL PERFORMANCE CHARACTERISTICS (Continued)

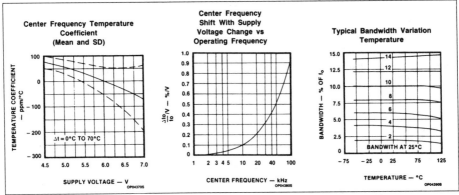

Center Frequency Temperature Coefficient (Mean and SD)

Center Frequency Shift With Supply Voltage Change vs Operating Frequency

Typical Bandwidth Variation Temperature

DESIGN FORMULAS

$$f_O \cong \frac{1}{1.1 R_1 C_1}$$

$$BW \simeq 1070 \sqrt{\frac{V_I}{f_O C_2}} \text{ in } \% \text{ of } f_O,$$

$$V_I \leqslant 200 mV_{RMS}$$

Where

V_I = Input voltage (V_{RMS})
C_2 = Low-pass filter capacitor (μF)

PHASE-LOCKED LOOP TERMINOLOGY CENTER FREQUENCY (f_O)

The free-running frequency of the current controlled oscillator (CCO) in the absence of an input signal.

Detection Bandwidth (BW)

The frequency range, centered about f_O, within which an input signal above the threshold voltage (typically 20mV$_{RMS}$) will cause a logical zero state on the output. The detection bandwidth corresponds to the loop capture range.

Lock Range

The largest frequency range within which an input signal above the threshold voltage will hold a logical zero state on the output.

Detection Band Skew

A measure of how well the detection band is centered about the center frequency, f_O. The skew is defined as $(f_{MAX} + f_{MIN} - 2f_O)/2f_O$ where fmax and fmin are the frequencies corresponding to the edges of the detection band. The skew can be reduced to zero if necessary by means of an optional centering adjustment.

OPERATING INSTRUCTIONS

Figure 1 shows a typical connection diagram for the 567. For most applications, the following three-step procedure will be sufficient for choosing the external components R_1, C_1, C_2 and C_3.

1. Select R_1 and C_1 for the desired center frequency. For best temperature stability, R_1 should be between 2K and 20K ohm, and the combined temperature coefficient of the $R_1 C_1$ product should have sufficient stability over the projected temperature range to meet the necessary requirements.

TYPICAL RESPONSE

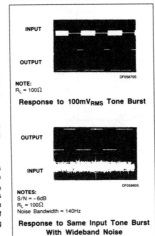

INPUT

OUTPUT

NOTE:
$R_L = 100\Omega$

Response to 100mV$_{RMS}$ Tone Burst

OUTPUT

INPUT

NOTES:
S/N = −6dB
$R_L = 100\Omega$
Noise Bandwidth = 140Hz

Response to Same Input Tone Burst With Wideband Noise

2. Select the low-pass capacitor, C_2, by referring to the Bandwidth versus Input Signal Amplitude graph. If the input amplitude Variation is known, the appropriate value of $f_O C_2$ necessary to give the desired bandwidth may be found. Conversely, an area of operation may be selected on this graph and the input level and C_2 may be adjusted accordingly. For example, constand bandwidth operation requires that input amplitude be above 200mVrms. The bandwidth, as noted on the graph, is then controlled solely by the $f_O C_2$ product (f_O (Hz), $C_2(\mu F)$).

3. The value of C_3 is generally non-critical. C_3 sets the band edge of a low-pass filter which attenuates frequencies outside the detection band to eliminate spurious outputs. If C_3 is too small, frequencies just outside the detection band will switch the output stage on and off at the beat frequency, or the output may pulse on and off during the turn-on transient. If C_3 is too large, turn-on and turn-off of the

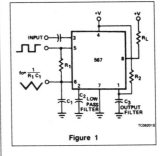

$f_O = \frac{1}{R_1 C_1}$

Figure 1

output stage will be delayed until the voltage on C_3 passes the threshold voltage. (Such delay may be desirable to avoid spurious outputs due to transient frequencies.) A typical minimum value for C_3 is $2C_2$.

4. Optional resistor R_2 sets the threshold for the largest "no output" input voltage. A value of $130k\Omega$ is used to assure the tested limit of $10mV_{RMS}$ min. This resistor can be referenced to ground for increased sensitivity. The explanation can be found in the "optional controls" section which follows.

AVAILABLE OUTPUTS (Figure 2)

The primary output is the uncommitted output transistor collector, Pin 8. When an in-band input signal is present, this transistor saturates; its collector voltage being less than 1.0 volt (typically 0.6V) at full output current (100mA). The voltage at Pin 2 is the phase detector output which is a linear function of frequency over the range of 0.95 to 1.05 f_O with a slope of about 20mV per percent of frequency deviation. The average voltage at Pin 1 is, during lock, a function of the in-band input amplitude in accordance with the transfer characteristic given. Pin 5 is the controlled oscillator square wave output of magnitude $(+V -2V_{BE}) \approx (+V-1.4V)$ having a DC average of $+V/2$. A $1k\Omega$ load may be driven from pin 5. Pin 6 is an exponential triangle of $1V_{P-P}$ with an average DC level of $+V/2$. Only high impedance loads may be connected to pin 6 without affecting the CCO duty cycle or temperature stability.

OPERATING PRECAUTIONS

A brief review of the following precautions will help the user achieve the high level of performance of which the 567 is capable.

1. Operation in the high input level mode (above 200mV) will free the user from bandwidth variations due to changes in the in-band signal amplitude. The input stage is now limiting, however, so that out-band signals or high noise levels can cause an apparent bandwidth reduction as the inband signal is suppressed. Also, the limiting action will create in-band components from sub-harmonic signals, so the 567 becomes sensitive to signals at $f_O/3$, $f_O/5$, etc.

2. The 567 will lock onto signals near $(2n + 1)$ f_O, and will give an output for signals near $(4n + 1)$ f_O where $n = 0, 1, 2$, etc. Thus, signals at $5f_O$ and $9f_O$ can cause an unwanted output. If such signals are anticipated, they should be attenuated before reaching the 567 input.

3. Maximum immunity from noise and outband signals is afforded in the low input

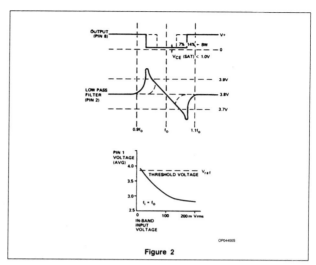

Figure 2

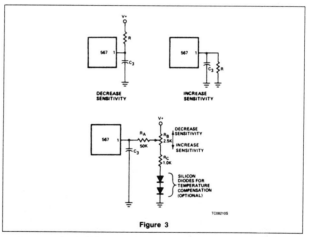

Figure 3

level (below 200mV$_{RMS}$) and reduced bandwidth operating mode. However, decreased loop damping causes the worst-case lock-up time to increase, as shown by the Greatest Number of Cycles Before Output vs Bandwidth graph.

4. Due to the high switching speeds (20ns) associated with 567 operation, care should be taken in lead routing. Lead lengths should be kept to a minimum.

The power supply should be adequately bypassed close to the 567 with a $0.01\mu F$ or greater capacitor; grounding paths should be carefully chosen to avoid ground loops and unwanted voltage variations. Another factor which must be considered is the effect of load energization on the power supply. For example, an incandescent lamp typically draws 10 times rated current at turn-on. This can

173

cause supply voltage fluctuations which could, for example, shift the detection band of narrow-band systems sufficiently to cause momentary loss of lock. The result is a low-frequency oscillation into and out of lock. Such effects can be prevented by supplying heavy load currents from a separate supply or increasing the supply filter capacitor.

SPEED OF OPERATION

Minimum lock-up time is related to the natural frequency of the loop. The lower it is, the longer becomes the turn-on transient. Thus, maximum operating speed is obtained when C_2 is at a minimum. When the signal is first applied, the phase may be such as to initially drive the controlled oscillator away from the incoming frequency rather than toward it. Under this condition, which is of course unpredictable, the lock-up transient is at its worst and the theoretical minimum lock-up time is not achievable. We must simply wait for the transient to die out.

The following expressions give the values of C_2 and C_3 which allow highest operating speeds for various band center frequencies. The minimum rate at which digital information may be detected without information loss due to the turn-on transient or output chatter is about 10 cycles per bit, corresponding to an information transfer rate of $f_O/10$ baud.

$$C_2 = \frac{130}{f_O} \mu F$$

$$C_3 = \frac{260}{f_O} \mu F$$

In cases where turn-off time can be sacrificed to achieve fast turn-on, the optional sensitivity adjustment circuit can be used to move the quiescent C_3 voltage lower (closer to the threshold voltage). However, sensitivity to beat frequencies, noise and extraneous signals will be increased.

OPTIONAL CONTROLS (Figure 3)

The 567 has been designed so that, for most applications, no external adjustments are required. Certain applications, however, will be greatly facilitated if full advantage is taken of the added control possibilities available through the use of additional external components. In the diagrams given, typical values are suggested where applicable. For best results the resistors used, except where noted, should have the same temperature coefficient. Ideally, silicon diodes would be low-resistivity types, such as forward-biased tran-

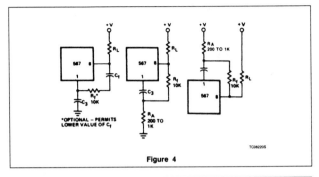

Figure 4

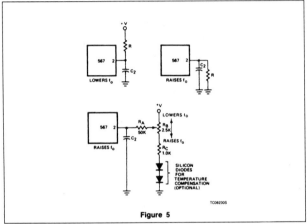

Figure 5

sistor base-emitter junctions. However, ordinary low-voltage diodes should be adequate for most applications.

SENSITIVITY ADJUSTMENT

(Figure 3)

When operated as a very narrow-band detector (less than 8 percent), both C_2 and C_3 are made quite large in order to improve noise and out-band signal rejection. This will inevitably slow the response time. If, however, the output stage is biased closer to the threshold level, the turn-on time can be improved. This is accomplished by drawing additional current to terminal 1. Under this condition, the 567

will also give an output for lower-level signals (10mV or lower).

By adding current to terminal 1, the output stage is biased further away from the threshold voltage. This is most useful when, to obtain maximum operating speed, C_2 and C_3 are made very small. Normally, frequencies just outside the detection band could cause false outputs under this condition. By desensitizing the output stage, the out-band beat notes do not feed through to the output stage. Since the input level must be somewhat greater when the output stage is made less sensitive, rejection of third harmonics or in-band harmonics (of lower frequency signals) is also improved.

TYPICAL APPLICATIONS

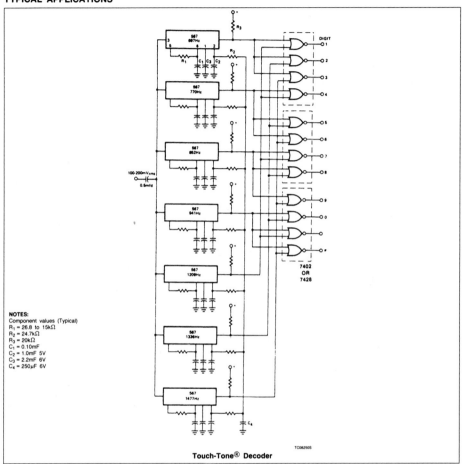

NOTES:
Component values (Typical)
R_1 = 26.8 to 15kΩ
R_2 = 24.7kΩ
R_3 = 20kΩ
C_1 = 0.10mF
C_2 = 1.0mF 5V
C_3 = 2.2mF 6V
C_4 = 250µF 6V

Touch-Tone® Decoder

TYPICAL APPLICATIONS (Continued)

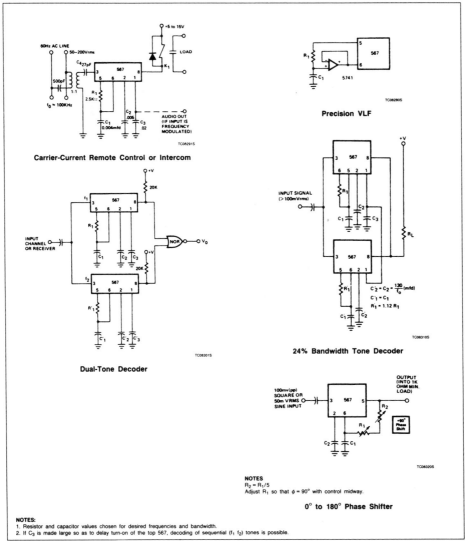

Carrier-Current Remote Control or Intercom

Precision VLF

Dual-Tone Decoder

24% Bandwidth Tone Decoder

0° to 180° Phase Shifter

NOTES
$R_2 = R_1/5$
Adjust R_1 so that $\phi = 90°$ with control midway.

NOTES:
1. Resistor and capacitor values chosen for desired frequencies and bandwidth.
2. If C_3 is made large so as to delay turn-on of the top 567, decoding of sequential (f_1 f_2) tones is possible.

SSI 75T202/203
5V Low-Power
DTMF Receiver

Data Sheet

DESCRIPTION

The SSI 202 and 203 are complete Dual Tone Multiple Frequency (DTMF) receivers detecting a selectable group of 12 or 16 standard digits. No front-end pre-filtering is needed. The only externally required components are an inexpensive 3.58-MHz television "colorburst" crystal (for frequency reference) and a bias resistor. Extremely high system density is made possible by using the clock output of a crystal connected SSI 202 or 203 receiver to drive the time bases of additional receivers. Both are monolithic integrated circuits fabricated with low-power, complementary symmetry MOS (CMOS) processing. They require only a single low tolerance voltage supply and are packaged in a standard 18 pin plastic DIP.

The SSI 202 and 203 employ state-of-the-art circuit technology to combine digital and analog functions on the same CMOS chip using a standard digital semiconductor process. The analog input is pre-processed by 60-Hz reject and band splitting filters and then hard-limited to provide AGC. Eight bandpass filters detect the individual

tones. The digital post-processor times the tone durations and provides the correctly coded digital outputs. Outputs interface directly to standard CMOS circuitry, and are three-state enabled to facilitate bus-oriented architectures.

FEATURES

- **Central office quality**
- **NO front-end band-splitting filters required**
- **Single, low-tolerance, 5-volt supply**
- **Detects either 12 or 16 standard DTMF digits**
- **Uses inexpensive 3.579545-MHz crystal for reference**
- **Excellent speech immunity**
- **Output in either 4-bit hexadecimal code or binary coded 2 of 8**
- **18-pin DIP package for high system density**
- **Synchronous or handshake interface**
- **Three-state outputs**
- **Early detect output (SSI 203 only)**

SSI 202/203 Block Diagram

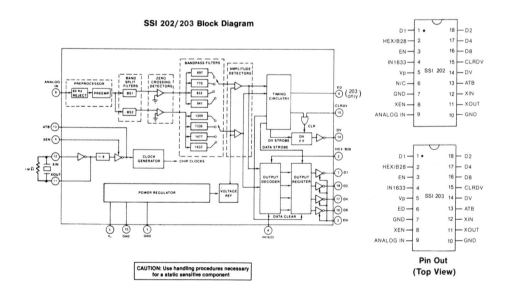

CAUTION: Use handling procedures necessary for a static sensitive component

**Pin Out
(Top View)**

177

SSI 75T202/203
5V Low-Power
DTMF Receiver

ANALOG IN
This pin accepts the analog input. It is internally biased so that the input signal may be AC coupled. The input may be DC coupled as long as it does not exceed the positive supply. Proper input coupling is illustrated below.

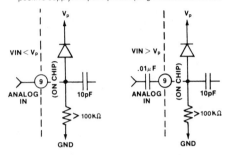

The SSI 202 is designed to accept sinusoidal input wave forms but will operate satisfactorily with any input that has the correct fundamental frequency with harmonics less than -20 dB below the fundamental.

CRYSTAL OSCILLATOR
The SSI 202 and 203 contain an onboard inverter with sufficient gain to provide oscillation when connected to a low-cost television "color-burst" crystal. The crystal oscillator is enabled by tying XEN high. The crystal is connected between XIN and XOUT. A 1 MΩ 10% resistor is also connected between these pins. In this mode, ATB is a clock frequency output. Other SSI 202's (or 203's) may use the same frequency reference by tying their ATB pins to the ATB of a crystal connected device. XIN and XEN of the auxiliary devices must then be tied high and low respectively. Ten devices may run off a single crystal-connected SSI 202 or 203 as shown below.

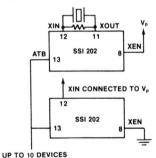

UP TO 10 DEVICES

HEX/B28
This pin selects the format of the digital output code. When HEX/B28 is tied high, the output is hexadecimal. When tied low, the output is binary coded 2 of 8. The table below describes the two output codes.

Digit	Hexadecimal				Binary Coded 2 of 8			
	D8	D4	D2	D1	D8	D4	D2	D1
1	0	0	0	1	0	0	0	0
2	0	0	1	0	0	0	0	1
3	0	0	1	1	0	0	1	0
4	0	1	0	0	0	1	0	0
5	0	1	0	1	0	1	0	1
6	0	1	1	0	0	1	1	0
7	0	1	1	1	1	0	0	0
8	1	0	0	0	1	0	0	1
9	1	0	0	1	1	0	1	0
0	1	0	1	0	1	1	0	1
*	1	0	1	1	1	1	0	0
#	1	1	0	0	1	1	1	0
A	1	1	0	1	0	0	1	1
B	1	1	1	0	0	1	1	1
C	1	1	1	1	1	0	1	1
D	0	0	0	0	1	1	1	1

IN1633
When tied high, this pin inhibits detection of tone pairs containing the 1633-Hz component. For detection of all 16 standard digits, IN1633 must be tied low.

OUTPUTS D1, D2, D4, D8 and EN
Outputs D1, D2, D4, D8 are CMOS push-pull when enabled (EN high) and open circuited (high impedence) when disabled by pulling EN low. These digital outputs provide the code corresponding to the detected digit in the format programmed by the HEX/B28 pin. The digital outputs become valid after a tone pair has been detected and they are then cleared when a valid pause is timed.

DV and CLRDV

DV signals a detection by going high after a valid tone pair is sensed and decoded at the output pins D1, D2, D4, D8. DV remains high until a valid pause occurs or the CLRDV is raised high, whichever is earlier.

ED (SSI 203 only)

The ED output goes high as soon as the SSI 203 begins to detect a DTMF tone pair and falls when the 203 begins to detect a pause. The D1, D2, D4, and D8 outputs are guaranteed to be valid when DV is high, but are not necessarily valid when ED is high.

N/C PINS

These pins have no internal connection and may be left floating.

SSI 202/203 TIMING

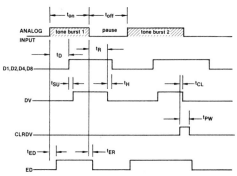

DTMF DIALING MATRIX

	Col 0	Col 1	Col 2	Col 3
Row 0	1	2	3	A
Row 1	4	5	6	B
Row 2	7	8	9	C
Row 3	*	0	#	D

Note: Column 3 is for special applications and is not normally used in telephone dialing.

DETECTION FREQUENCY

Low Group f_o	High Group f_o
Row 0 = 697 Hz	Column 0 = 1209 Hz
Row 1 = 770 Hz	Column 1 = 1336 Hz
Row 2 = 852 Hz	Column 2 = 1477 Hz
Row 3 = 941 Hz	Column 3 = 1633 Hz

PARAMETER	SYMBOL	MIN.	NOM.	MAX.	UNITS
TONE TIME: for detection	t_{ON}	40	—	—	ms
for rejection	t_{ON}	—	—	20	ms
PAUSE TIME: for detection	t_{OFF}	40	—	—	ms
for rejection	t_{OFF}	—	—	20	ms
DETECT TIME	t_D	25	—	46	ms
RELEASE TIME	t_R	35	—	50	ms
DATA SETUP TIME	t_{SU}	7	—	—	µS
DATA HOLD TIME	t_H	4.2	—	5.0	ms
DV CLEAR TIME	t_{CL}	—	160	250	ns
CLRDV pulse width	t_{PW}	200	—	—	ns
ED Detect Time	t_{ED}	7	—	22	ms
ED Release Time	t_{ER}	2	—	18	ms
OUTPUT ENABLE TIME $C_L = 50pF \ R_L = 1K\Omega$	—	—	—	200	ns
OUTPUT DISABLE TIME $C_L = 35pF \ R_L = 500\Omega$	—	—	—	200	ns
OUTPUT RISE TIME $C_L = 50pF$	—	—	—	200	ns
OUTPUT FALL TIME $C_L = 50pF$	—	—	160	200	ns

silicon systems™

14351 Myford Road, Tustin, CA 92680   (714) 731-7110, TWX 910-595-2809

ABSOLUTE MAXIMUM RATINGS*

DC Supply Voltage V_p . . . . +7 V
Operating Temperature . . . . . . -40°C to +85°C Ambient
Storage Temperature . . . . -65°C to 150°C
Power Dissipation (25°C) . . . . 65 mW
 (Derate above T_A = 25°C @ 6.25 mW/°C)

Input Voltage . . . . $(V_p + .5V)$ to – .5V
 (All inputs except ANALOG IN)
ANALOG IN Voltage . . . . $(V_p + .5V)$ to $(V_p - 10V)$
DC Current into any Input . . . . ±1.0mA
Lead Temperature . . . . 300°C
 (soldering, 10 sec.)

*Operation above absolute maximum ratings may damage the device.
Note: All SSI 202/203 unused inputs must be connected to V_p or Gnd,
as appropriate.

ELECTRICAL CHARACTERISTICS ($-40°C \leqslant T_A \leqslant +85°C$, V_p = 5V ± 10%)

Parameter	Conditions	Min	Typ	Max	Units
Frequency Detect Bandwidth		±(1.5 + 2 Hz)	±2.3	± 3.5	% of f_0
Amplitude for Detection	each tone	-32		-2	dBm referenced to 600 Ω
Minimum Acceptable Twist	twist = $\frac{\text{high tone}}{\text{low tone}}$	-10		+10	dB
60-Hz Tolerance				0.8	Vrms
Dial Tone Tolerance	"precise" dial tone			0dB	dB referenced to lower amplitude tone
Talk Off	MITEL tape #CM 7290		2		hits
Digital Outputs (except XOUT)	"0" level, 400 μA load "1" level, 200 μA load	0 V_p - 0.5		0.5 V_p	V V
Digital Inputs	"0" level "1" level	0 0.7 V_p)		0.3 V_L V_L	V V
Power Supply Noise	wide band			10	mV p-p
Supply Current	T_A = 25°C		10	16	mA
Noise Tolerance	MITEL tape #CM 7290			-12	dB referenced to lowest amplitude tone
Input Impedence	$V_p \geqslant V_{in} \geqslant V_p$-10	100 KΩ // 15pF			

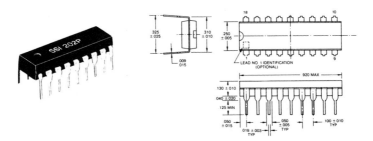

 **MOTOROLA**

PLL FREQUENCY SYNTHESIZER

The MC145106 is a phase locked loop (PLL) frequency synthesizer constructed in CMOS on a single monolithic structure. This synthesizer finds applications in such areas as CB and FM transceivers. The device contains an oscillator/amplifier, a 2^{10} or 2^{11} divider chain for the oscillator signal, a programmable divider chain for the input signal and a phase detector. The MC145106 has circuitry for a 10.24 MHz oscillator or may operate with an external signal. The circuit provides a 5.12 MHz output signal, which can be used for frequency tripling. A 2^9 programmable divider divides the input signal frequency for channel selection. The inputs to the programmable divider are standard ground-to-supply binary signals. Pull-down resistors on these inputs normally set these inputs to ground enabling these programmable inputs to be controlled from a mechanical switch or electronic circuitry.

The phase detector may control a VCO and yields a high level signal when input frequency is low, and a low level signal when input frequency is high. An out of lock signal is provided from the on-chip lock detector with a "0" level for the out of lock condition.

- Single Power Supply
- Wide Supply Range: 4.5 to 12 V
- Provision for 10.24 MHz Crystal Oscillator
- 5.12 MHz Output
- Programmable Division Binary Input Selects up to 2^9
- On-Chip Pull Down Resistors on Programmable Divider Inputs
- Selectable Reference Divider, 2^{10} or 2^{11} (including ÷ 2)
- Three-State Phase Detector
- Pin-for-Pin Replacement for MM55106, MM55116
- Chip Complexity: 880 FETs or 220 Equivalent Gates

CMOS MSI

(LOW-POWER COMPLEMENTARY MOS)

PLL
FREQUENCY SYNTHESIZER

P SUFFIX
PLASTIC PACKAGE
CASE 707

BLOCK DIAGRAM

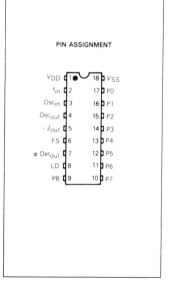

PIN ASSIGNMENT

MC145106

MAXIMUM RATINGS (Voltages referenced to V_{SS})

Rating	Symbol	Value	Unit
DC Supply Voltage	V_{DD}	−0.5 to +12	V
Input Voltage, All Inputs	V_{in}	−0.5 to V_{DD} +0.5	V
DC Input Current, per Pin	I	±10	mA
Operating Temperature Range	T_A	−40 to +85	°C
Storage Temperature Range	T_{stg}	−65 to +150	°C

ELECTRICAL CHARACTERISTICS

($T_A = 25°C$ Unless Otherwise Stated, Voltages Referenced to V_{SS})

Characteristic	Symbol	V_{DD} Vdc	All Types Min	Typ	Max	Unit
Power Supply Voltage Range	V_{DD}	—	4.5	—	12	V
Supply Current	I_{DD}	5.0	—	6	10	mA
		10	—	20	35	
		12	—	28	50	
Input Voltage "0" Level	V_{IL}	5.0	—	—	1.5	V
		10	—	—	3.0	
		12	—	—	3.6	
"1" Level	V_{IH}	5.0	3.5	—	—	
		10	7.0	—	—	
		12	8.4	—	—	
Input Current "0" Level (FS, Pull up Resistor Source Current)	I_{in}	5.0	−5.0	−20	−50	µA
		10	−15	−60	−150	
		12	−20	−80	−200	
(P0 to P8)		5.0	—	—	−0.3	
		10	—	—	−0.3	
		12	—	—	−0.3	
(FS) "1" Level		5.0	—	—	0.3	
		10	—	—	0.3	
		12	—	—	0.3	
(P0 to P8, Pull down Resistor Sink Current)		5.0	7.5	30	75	
		10	22.5	90	225	
		12	30	120	300	
(Osc_{in}, f_{in}) "0" Level		5.0	−2.0	−6.0	−15	
		10	−6.0	−25	−62	
		12	−9.0	−37	−92	
(Osc_{in}, f_{in}) "1" Level		5.0	2.0	6.0	15	
		10	6.0	25	62	
		12	9.0	37	92	
Output Drive Current Source	I_{OH}					mA
($V_O = 4.5$ V)		5.0	−0.7	−1.4	—	
($V_O = 9.5$ V)		10	−1.1	−2.2	—	
($V_O = 11.5$ V)		12	−1.5	−3.0	—	
($V_O = 0.5$ V) Sink	I_{OL}	5.0	0.9	1.8	—	
($V_O = 0.5$ V)		10	1.4	2.8	—	
($V_O = 0.5$ V)		12	2.0	4.0	—	
Input Amplitude	—					Vp-p Sine
(f_{in} @ 4.0 MHz)		—	1.0	0.2	—	
(Osc_{in} @10.24 MHz)		—	1.5	0.3	—	
Input Resistance (Osc_{in}, f_{in})	P_{in}					MΩ
		5.0	—	1.0	—	
		10	—	0.5	—	
		12	—	—	—	
Input Capacitance (Osc_{in}, f_{in})	C_{in}	—	—	6.0		pF
Three State Leakage Current (φ Det$_{out}$)	I_{OZ}	5.0	—	—	1.0	µA
		10	—	—	1.0	
		12	—	—	1.0	
Input Frequency (−40°C to +85°C)	f_{in}	4.5	0	—	4.0	MHz
		12	0	—	4.0	
Oscillator Frequency (−40°C to +85°C)	Osc_{in}	4.5	0.1	—	10.24	MHz
		12	0.1	—	10.24	

Copyright of Motorola, Inc. Used by Permission.

MC145106

FIGURE 1 – MAXIMUM DIVIDER INPUT FREQUENCY versus SUPPLY VOLTAGE

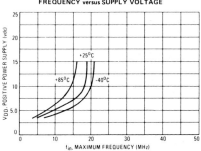

FIGURE 2 – MAXIMUM OSCILLATOR INPUT FREQUENCY versus SUPPLY VOLTAGE

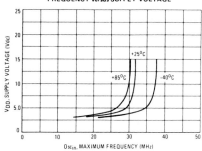

TRUTH TABLE

P8	P7	P6	P5	P4	P3	P2	P1	P0	Divide By N
				Selection					
0	0	0	0	0	0	0	0	0	2 (Note 1)
0	0	0	0	0	0	0	0	1	3 (Note 1)
0	0	0	0	0	0	0	1	0	2
0	0	0	0	0	0	0	1	1	3
0	0	0	0	0	0	1	0	0	4
.	.	.	.	.	.	.	.	.	.
.	.	.	.	.	.	.	.	.	.
0	1	1	1	1	1	1	1	1	255
.	.	.	.	.	.	.	.	.	.
.	.	.	.	.	.	.	.	.	.
1	1	1	1	1	1	1	1	1	511

1: Voltage level = V_{DD}

0: Voltage level = 0 or open circuit input

Note 1: The binary setting of 00000000 and 00000001 on P8 to P0 results in a 2 and 3 division which is not in the 2^N-1 sequence. When pin is not connected the logic signal on that pin can be treated as a "0".

PIN DESCRIPTIONS

P0 – P8 – Programmable divider inputs (binary)

f_{in} – Frequency input to programmable divider (derived from VCO)

Osc_{in} – Oscillator/amplifier input terminal

Osc_{out} – Oscillator/amplifier output terminal

LD – Lock detector, high when loop is locked, pulses low when out of lock.

$\phi \, Det_{out}$ – Signal for control of external VCO, output high when f_{in}/N is less than the reference frequency; output low when f_{in}/N is greater than the reference frequency. Reference frequency is the divided down oscillator - input frequency typically 5.0 or 10 kHz.

FS – Reference Oscillator Frequency Division Select. When using 10.24 MHz Osc frequency, this control selects 10 kHz, a "0" selects 5.0 kHz.

$\div 2_{out}$ – Reference Osc frequency divided by 2 output; when using 10.24 MHz Osc frequency, this output is 5.12 MHz for frequency tripling applications.

V_{DD} – Positive power supply

V_{SS} – Ground

National Semiconductor

ADC0808/ADC0809 8-Bit μP Compatible A/D Converters with 8-Channel Multiplexer

General Description

The ADC0808, ADC0809 data acquisition component is a monolithic CMOS device with an 8-bit analog-to-digital converter, 8-channel multiplexer and microprocessor compatible control logic. The 8-bit A/D converter uses successive approximation as the conversion technique. The converter features a high impedance chopper stabilized comparator, a 256R voltage divider with analog switch tree and a successive approximation register. The 8-channel multiplexer can directly access any of 8-single-ended analog signals.

The device eliminates the need for external zero and full-scale adjustments. Easy interfacing to microprocessors is provided by the latched and decoded multiplexer address inputs and latched TTL TRI-STATE® outputs.

The design of the ADC0808, ADC0809 has been optimized by incorporating the most desirable aspects of several A/D conversion techniques. The ADC0808, ADC0809 offers high speed, high accuracy, minimal temperature dependence, excellent long-term accuracy and repeatability, and consumes minimal power. These features make this device ideally suited to applications from process and machine control to consumer and automotive applications. For 16-channel multiplexer with common output (sample/hold port) see ADC0816 data sheet. (See AN-247 for more information.)

Features

- Easy interface to all microprocessors
- Operates ratiometrically or with 5 V_{DC} or analog span adjusted voltage reference
- No zero or full-scale adjust required
- 8-channel multiplexer with address logic
- 0V to 5V input range with single 5V power supply
- Outputs meet TTL voltage level specifications
- Standard hermetic or molded 28-pin DIP package
- 28-pin molded chip carrier package
- ADC0808 equivalent to MM74C949
- ADC0809 equivalent to MM74C949-1

Key Specifications

■ Resolution	8 Bits
■ Total Unadjusted Error	± ½ LSB and ±1 LSB
■ Single Supply	5 V_{DC}
■ Low Power	15 mW
■ Conversion Time	100 μs

Block Diagram

TL/H/5672-1

National Semiconductor Corporation	National Semiconductor GmbH	NS Japan Ltd.	National Semiconductor Hong Kong Ltd.	National Semicondutores Do Brasil Ltda.	National Semiconductor (Australia) PTY, Ltd.
2900 Semiconductor Drive	Westendstrasse 193-195	Sanseido Bldg. 5F	Southeast Asia Marketing	Av. Brig. Faria Lima, 830	21/3 High Street
P.O. Box 58090	D-8000 Munchen 21	4-15 Nishi Shinjuku	Austin Tower, 4th Floor	8 Andar	Bayswater, Victoria 3153
Santa Clara, CA 95052-8090	West Germany	Shinjuku-Ku,	22-26A Austin Avenue	01452 Sao Paulo, SP. Brasil	Australia
Tel: (408) 721-5000	Tel: (089) 5 70 95 01	Tokyo 160, Japan	Tsimshatsui, Kowloon, H.K.	Tel: (55/11) 212-5066	Tel: (03) 729-6333
TWX: (910) 339-9240	Telex: 522772	Tel: 3-299-7001	Tel: 3-7231290, 3-7243645	Telex: 391-1131931 NSBR BR	Telex: AA32096
		FAX: 3-299-7000	Cable: NSSEAMKTG		
			Telex: 52996 NSSEA HX		

National does not assume any responsibility for use of any circuitry described, no circuit patent licenses are implied and National reserves the right at any time without notice to change said circuitry and specifications.

Absolute Maximum Ratings (Notes 1 & 2)

If Military/Aerospace specified devices are required, please contact the National Semiconductor Sales Office/Distributors for availability and specifications.

Supply Voltage (V_{CC}) (Note 3)	6.5V
Voltage at Any Pin	$-0.3V$ to ($V_{CC}+0.3V$)
Except Control Inputs	
Voltage at Control Inputs	$-0.3V$ to $+15V$
(START, OE, CLOCK, ALE, ADD A, ADD B, ADD C)	
Storage Temperature Range	$-65°C$ to $+150°C$
Package Dissipation at $T_A = 25°C$	875 mW
Lead Temp. (Soldering, 10 seconds)	
Dual-In-Line Package (plastic)	260°C
Dual-In-Line Package (ceramic)	300°C
Molded Chip Carrier Package	
Vapor Phase (60 seconds)	215°C
Infrared (15 seconds)	220°C
ESD Susceptibility (Note 11)	400V

Operating Conditions (Notes 1 & 2)

Temperature Range (Note 1)	$T_{MIN} \leq T_A \leq T_{MAX}$
ADC0808CJ	$-55°C \leq T_A \leq +125°C$
ADC0808CCJ, ADC0808CCN,	
ADC0809CCN	$-40°C \leq T_A \leq +85°C$
ADC0808CCV, ADC0809CCV	$-40°C \leq T_A \leq +85°C$
Range of V_{CC} (Note 1)	4.5 V_{DC} to 6.0 V_{DC}

Electrical Characteristics

Converter Specifications: $V_{CC} = 5\ V_{DC} = V_{REF+}$, $V_{REF(-)} = GND$, $T_{MIN} \leq T_A \leq T_{MAX}$ and $f_{CLK} = 640$ kHz unless otherwise stated.

Symbol	Parameter	Conditions	Min	Typ	Max	Units
	ADC0808					
	Total Unadjusted Error	25°C			$\pm\frac{1}{2}$	LSB
	(Note 5)	T_{MIN} to T_{MAX}			$\pm\frac{3}{4}$	LSB
	ADC0809					
	Total Unadjusted Error	0°C to 70°C			±1	LSB
	(Note 5)	T_{MIN} to T_{MAX}			$\pm1\frac{1}{4}$	LSB
	Input Resistance	From Ref(+) to Ref(−)	1.0	2.5		kΩ
	Analog Input Voltage Range	(Note 4) V(+) or V(−)	$GND-0.10$		$V_{CC}+0.10$	V_{DC}
$V_{REF(+)}$	Voltage, Top of Ladder	Measured at Ref(+)		V_{CC}	$V_{CC}+0.1$	V
$\frac{V_{REF(+)} + V_{REF(-)}}{2}$	Voltage, Center of Ladder		$V_{CC}/2-0.1$	$V_{CC}/2$	$V_{CC}/2+0.1$	V
$V_{REF(-)}$	Voltage, Bottom of Ladder	Measured at Ref(−)	-0.1	0		V
I_{IN}	Comparator Input Current	$f_c = 640$ kHz, (Note 6)	-2	±0.5	2	μA

Electrical Characteristics

Digital Levels and DC Specifications: ADC0808CJ $4.5V \leq V_{CC} \leq 5.5V$, $-55°C \leq T_A \leq +125°C$ unless otherwise noted ADC0808CCJ, ADC0808CCN, ADC0808CCV, ADC0809CCN and ADC0809CCV, $4.75 \leq V_{CC} \leq 5.25V$, $-40°C \leq T_A \leq +85°C$ unless otherwise noted

Symbol	Parameter	Conditions	Min	Typ	Max	Units
ANALOG MULTIPLEXER						
$I_{OFF(+)}$	OFF Channel Leakage Current	$V_{CC} = 5V$, $V_{IN} = 5V$,				
		$T_A = 25°C$		10	200	nA
		T_{MIN} to T_{MAX}			1.0	μA
$I_{OFF(-)}$	OFF Channel Leakage Current	$V_{CC} = 5V$, $V_{IN} = 0$,				
		$T_A = 25°C$	-200	-10		nA
		T_{MIN} to T_{MAX}	-1.0			μA

National Semiconductor Corporation	National Semiconductor GmbH	NS Japan Ltd.	National Semiconductor Hong Kong Ltd.	National Semicondutores Do Brasil Ltda.	National Semiconductor (Australia) PTY, Ltd.
2900 Semiconductor Drive	Westendstrasse 193-195	Sanseido Bldg. 5F	Southeast Asia Marketing	Av. Brig. Faria Lima, 830	21/3 High Street
P.O. Box 58090	D-8000 Munchen 21	4-15 Nishi Shinjuku	Austin Tower, 4th Floor	8 Andar	Bayswater, Victoria 3153
Santa Clara, CA 95052-8090	West Germany	Shinjuku-Ku,	22-26A Austin Avenue	01452 Sao Paulo, SP. Brasil	Australia
Tel: (408) 721-5000	Tel: (089) 5 70 95 01	Tokyo 160, Japan	Tsimshatsui, Kowloon, H.K.	Tel: (55/11) 212-5066	Tel: (03) 729-6333
TWX: (910) 339-9240	Telex: 522772	Tel: 3-299-7001	Tel: 3-7231290, 3-7243645	Telex: 391-1131931 NSBR BR	Telex: AA32096
		FAX: 3-299-7000	Cable: NSSEAMKTG		
			Telex: 52996 NSSEA HX		

National does not assume any responsibility for use of any circuitry described, no circuit patent licenses are implied and National reserves the right at any time without notice to change said circuitry and specifications.

185

Electrical Characteristics (Continued)

Digital Levels and DC Specifications: ADC0808CJ 4.5V ≤ V_{CC} ≤ 5.5V, −55°C ≤ T_A ≤ +125°C unless otherwise noted ADC0808CCJ, ADC0808CCN, ADC0808CCV, ADC0809CCN and ADC0809CCV, 4.75 ≤ V_{CC} ≤ 5.25V, −40°C ≤ T_A ≤ +85°C unless otherwise noted

Symbol	Parameter	Conditions	Min	Typ	Max	Units
CONTROL INPUTS						
$V_{IN(1)}$	Logical "1" Input Voltage		V_{CC} − 1.5			V
$V_{IN(0)}$	Logical "0" Input Voltage				1.5	V
$I_{IN(1)}$	Logical "1" Input Current (The Control Inputs)	V_{IN} = 15V			1.0	μA
$I_{IN(0)}$	Logical "0" Input Current (The Control Inputs)	V_{IN} = 0		−1.0		μA
I_{CC}	Supply Current	f_{CLK} = 640 kHz		0.3	3.0	mA
DATA OUTPUTS AND EOC (INTERRUPT)						
$V_{OUT(1)}$	Logical "1" Output Voltage	I_O = −360 μA	V_{CC} − 0.4			V
$V_{OUT(0)}$	Logical "0" Output Voltage	I_O = 1.6 mA			0.45	V
$V_{OUT(0)}$	Logical "0" Output Voltage EOC	I_O = 1.2 mA			0.45	V
I_{OUT}	TRI-STATE Output Current	V_O = 5V V_O = 0		−3	3	μA μA

Electrical Characteristics

Timing Specifications V_{CC} = $V_{REF(+)}$ = 5V, $V_{REF(−)}$ = GND, t_r = t_f = 20 ns and T_A = 25°C unless otherwise noted.

Symbol	Parameter	Conditions	Min	Typ	Max	Units
t_{WS}	Minimum Start Pulse Width	(Figure 5)		100	200	ns
t_{WALE}	Minimum ALE Pulse Width	(Figure 5)		100	200	ns
t_s	Minimum Address Set-Up Time	(Figure 5)		25	50	ns
t_H	Minimum Address Hold Time	(Figure 5)		25	50	ns
t_D	Analog MUX Delay Time From ALE	R_S = 0Ω (Figure 5)		1	2.5	μS
t_{H1}, t_{H0}	OE Control to Q Logic State	C_L = 50 pF, R_L = 10k (Figure 8)		125	250	ns
t_{1H}, t_{0H}	OE Control to Hi-Z	C_L = 10 pF, R_L = 10k (Figure 8)		125	250	ns
t_c	Conversion Time	f_c = 640 kHz, (Figure 5) (Note 7)	90	100	116	μS
f_c	Clock Frequency		10	640	1280	kHz
t_{EOC}	EOC Delay Time	(Figure 5)	0		8 + 2 μS	Clock Periods
C_{IN}	Input Capacitance	At Control Inputs		10	15	pF
C_{OUT}	TRI-STATE Output Capacitance	At TRI-STATE Outputs, (Note 12)		10	15	pF

Note 1: Absolute Maximum Ratings indicate limits beyond which damage to the device may occur. DC and AC electrical specifications do not apply when operating the device beyond its specified operating conditions.

Note 2: All voltages are measured with respect to GND, unless otherwise specified.

Note 3: A zener diode exists, internally, from V_{CC} to GND and has a typical breakdown voltage of 7 V_{DC}.

Note 4: Two on-chip diodes are tied to each analog input which will forward conduct for analog input voltages one diode drop below ground or one diode drop greater than the V_{CC} supply. The spec allows 100 mV forward bias of either diode. This means that as long as the analog V_{IN} does not exceed the supply voltage by more than 100 mV, the output code will be correct. To achieve an absolute 0V_{DC} to 5V_{DC} input voltage range will therefore require a minimum supply voltage of 4.900 V_{DC} over temperature variations, initial tolerance and loading.

Note 5: Total unadjusted error includes offset, full-scale, linearity, and multiplexer errors. See Figure 3. None of these A/Ds requires a zero or full-scale adjust. However, if an all zero code is desired for an analog input other than 0.0V, or if a narrow full-scale span exists (for example: 0.5V to 4.5V full-scale) the reference voltages can be adjusted to achieve this. See Figure 13.

Note 6: Comparator input current is a bias current into or out of the chopper stabilized comparator. The bias current varies directly with clock frequency and has little temperature dependence (Figure 6). See paragraph 4.0.

Note 7: The outputs of the data register are updated one clock cycle before the rising edge of EOC.

Note 8: Human body model, 100 pF discharged through a 1.5 kΩ resistor.

LIFE SUPPORT POLICY

NATIONAL'S PRODUCTS ARE NOT AUTHORIZED FOR USE AS CRITICAL COMPONENTS IN LIFE SUPPORT DEVICES OR SYSTEMS WITHOUT THE EXPRESS WRITTEN APPROVAL OF THE PRESIDENT OF NATIONAL SEMICONDUCTOR CORPORATION. As used herein:

1. Life support devices or systems are devices or systems which, (a) are intended for surgical implant into the body, or (b) support or sustain life, and whose failure to perform, when properly used in accordance with instructions for use provided in the labeling, can be reasonably expected to result in a significant injury to the user.

2. A critical component is any component of a life support device or system whose failure to perform can be reasonably expected to cause the failure of the life support device or system, or to affect its safety or effectiveness.

National Semiconductor Corporation	National Semiconductor GmbH	NS Japan Ltd.	National Semiconductor Hong Kong Ltd.	National Semicondutores Do Brasil Ltda.	National Semiconductor (Australia) PTY, Ltd.
2900 Semiconductor Drive P.O. Box 58090 Santa Clara, CA 95052-8090 Tel: (408) 721-5000 TWX: (910) 339-9240	Westendstrasse 193-195 D-8000 Munchen 21 West Germany Tel: (089) 5 70 95 01 Telex: 522772	Sanseido Bldg. 5F 4-15 Nishi Shinjuku Shinjuku-Ku, Tokyo 160, Japan Tel: 3-299-7001 FAX: 3-299-7000	Southeast Asia Marketing Austin Tower, 4th Floor 22-26A Austin Avenue Tsimshatsui, Kowloon, H.K. Tel: 3-7231290, 3-7243645 Cable: NSSEAMKTG Telex: 52996 NSSEA HX	Av. Brig. Faria Lima, 830 8 Andar 01452 Sao Paulo, SP, Brasil Tel: (55/11) 212-5066 Telex: 391-1131931 NSBR BR	21/3 High Street Bayswater, Victoria 3153 Australia Tel: (03) 729-6333 Telex: AA32096

National does not assume any responsibility for use of any circuitry described, no circuit patent licenses are implied and National reserves the right at any time without notice to change said circuitry and specifications.

186

ADC0808/ADC0809

Connection Diagrams

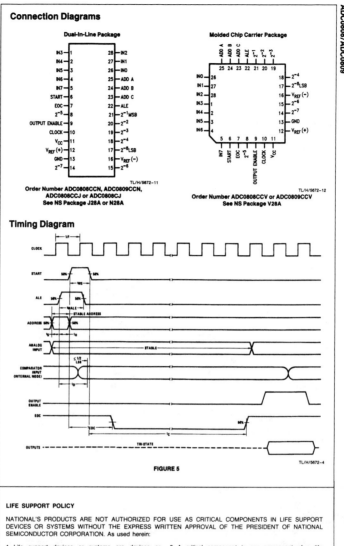

Dual-In-Line Package

Molded Chip Carrier Package

TL/H/5672–11

**Order Number ADC0808CCN, ADC0809CCN,
ADC0808CCJ or ADC0808CJ
See NS Package J28A or N28A**

TL/H/5672–12

**Order Number ADC0808CCV or ADC0809CCV
See NS Package V28A**

Timing Diagram

FIGURE 5

TL/H/5672–4

LIFE SUPPORT POLICY

NATIONAL'S PRODUCTS ARE NOT AUTHORIZED FOR USE AS CRITICAL COMPONENTS IN LIFE SUPPORT DEVICES OR SYSTEMS WITHOUT THE EXPRESS WRITTEN APPROVAL OF THE PRESIDENT OF NATIONAL SEMICONDUCTOR CORPORATION. As used herein:

1. Life support devices or systems are devices or systems which, (a) are intended for surgical implant into the body, or (b) support or sustain life, and whose failure to perform, when properly used in accordance with instructions for use provided in the labeling, can be reasonably expected to result in a significant injury to the user.

2. A critical component is any component of a life support device or system whose failure to perform can be reasonably expected to cause the failure of the life support device or system, or to affect its safety or effectiveness.

National Semiconductor Corporation
2900 Semiconductor Drive
P.O. Box 58090
Santa Clara, CA 95052-8090
Tel: (408) 721-5000
TWX: (910) 339-9240

National Semiconductor GmbH
Westendstrasse 193-195
D-8000 Munchen 21
West Germany
Tel: (089) 5 70 95 01
Telex: 522772

NS Japan Ltd.
Sanseido Bldg. 5F
4-15 Nishi Shinjuku
Shinjuku-Ku,
Tokyo 160, Japan
Tel: 3-299-7001
FAX: 3-299-7000

National Semiconductor Hong Kong Ltd.
Southeast Asia Marketing
Austin Tower, 4th Floor
22-26A Austin Avenue
Tsimshatsui, Kowloon, H.K.
Tel: 3-7231290, 3-7243645
Cable: NSSEAMKTG
Telex: 52996 NSSEA HX

National Semicondutores Do Brasil Ltda.
Av. Brig. Faria Lima, 830
8 Andar
01452 Sao Paulo, SP, Brasil
Tel: (55/11) 212-5066
Telex: 391-1131931 NSBR BR

National Semiconductor (Australia) PTY, Ltd.
21/3 High Street
Bayswater, Victoria 3153
Australia
Tel: (03) 729-6333
Telex: AA32096

National does not assume any responsibility for use of any circuitry described, no circuit patent licenses are implied and National reserves the right at any time without notice to change said circuitry and specifications.

Typical Application

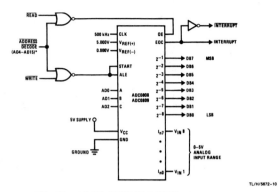

TL/H/5672–10

*Address latches needed for 8085 and SC/MP interfacing the ADC0808 to a microprocessor

MICROPROCESSOR INTERFACE TABLE

PROCESSOR	READ	WRITE	INTERRUPT (COMMENT)
8080	MEMR	MEMW	INTR (Thru RST Circuit)
8085	RD	WR	INTR (Thru RST Circuit)
Z-80	RD	WR	INT (Thru RST Circuit, Mode 0)
SC/MP	NRDS	NWDS	SA (Thru Sense A)
6800	VMA•φ2•R/W	VMA•φ•R/W	IRQA or IRQB (Thru PIA)

Ordering Information

TEMPERATURE RANGE		−40°C to +85°C			−55°C to +125°C
Error	± ½ LSB Unadjusted	ADC0808CCN	ADC0808CCV		ADC0808CCJ	ADC0808CJ
	± 1 LSB Unadjusted	ADC0809CCN	ADC0809CCV			
	Package Outline	N28A Molded DIP	V28A Molded Chip Carrier		J28A Ceramic DIP	J28A Ceramic DIP

National Semiconductor Corporation	National Semiconductor GmbH	NS Japan Ltd.	National Semiconductor Hong Kong Ltd.	National Semicondutores Do Brasil Ltda.	National Semiconductor (Australia) PTY, Ltd.
2900 Semiconductor Drive	Westendstrasse 193-195	Sanseido Bldg. 5F	Southeast Asia Marketing	Av. Brig. Faria Lima, 830	21/3 High Street
P.O. Box 58090	D-8000 Munchen 21	4-15 Nishi Shinjuku	Austin Tower, 4th Floor	8 Andar	Bayswater, Victoria 3153
Santa Clara, CA 95052-8090	West Germany	Shinjuku-Ku,	22-26A Austin Avenue	01452 Sao Paulo, SP, Brasil	Australia
Tel: (408) 721-5000	Tel: (089) 5 70 95 01	Tokyo 160, Japan	Tsimshatsui, Kowloon, H.K.	Tel: (55/11) 212-5066	Tel: (03) 729-6333
TWX: (910) 339-9240	Telex: 522772	Tel: 3-299-7001	Tel: 3-7231290, 3-7243645	Telex: 391-1131931 NSBR BR	Telex: AA32096
		FAX: 3-299-7000	Cable: NSSEAMKTG		
			Telex: 52996 NSSEA HX		

National does not assume any responsibility for use of any circuitry described, no circuit patent licenses are implied and National reserves the right at any time without notice to change said circuitry and specifications.

National Semiconductor

DAC0808/DAC0807/DAC0806 8-Bit D/A Converters

General Description

The DAC0808 series is an 8-bit monolithic digital-to-analog converter (DAC) featuring a full scale output current settling time of 150 ns while dissipating only 33 mW with ±5V supplies. No reference current (I_{REF}) trimming is required for most applications since the full scale output current is typically ±1 LSB of 255 I_{REF}/ 256. Relative accuracies of better than ±0.19% assure 8-bit monotonicity and linearity while zero level output current of less than 4 μA provides 8-bit zero accuracy for $I_{REF} \geq 2$ mA. The power supply currents of the DAC0808 series are independent of bit codes, and exhibits essentially constant device characteristics over the entire supply voltage range.

The DAC0808 will interface directly with popular TTL, DTL or CMOS logic levels, and is a direct replacement for the MC1508/MC1408. For higher speed applications, see DAC0800 data sheet.

Features

- Relative accuracy: ±0.19% error maximum (DAC0808)
- Full scale current match: ±1 LSB typ
- 7 and 6-bit accuracy available (DAC0807, DAC0806)
- Fast settling time: 150 ns typ
- Noninverting digital inputs are TTL and CMOS compatible
- High speed multiplying input slew rate: 8 mA/μs
- Power supply voltage range: ±4.5V to ±18V
- Low power consumption: 33 mW @ ±5V

Block and Connection Diagrams

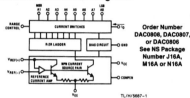

Order Number DAC0808, DAC0807, or DAC0806
See NS Package Number J16A, M16A or N16A

Dual-In-Line Package

TL/H/5687–1

TL/H/5687–2

Small-Outline Package

Top View

TL/H/5687–13

Ordering Information

ACCURACY	OPERATING TEMPERATURE RANGE	ORDER NUMBERS				
		J PACKAGE (J16A)*		N PACKAGE (N16A)*		SO PACKAGE (M16A)
8-bit	−55°C ≤ T_A ≤ +125°C	DAC0808LJ	MC1508L8			
8-bit	0°C ≤ T_A ≤ +75°C	DAC0808LCJ	MC1408L8	DAC0808LCN	MC1408P8	DAC0808LCM
7-bit	0°C ≤ T_A ≤ +75°C	DAC0807LCJ	MC1408L7	DAC0807LCN	MC1408P7	DAC0807LCM
6-bit	0°C ≤ T_A ≤ +75°C	DAC0806LCJ	MC1408L6	DAC0806LCN	MC1408P6	DAC0806LCM

*Note. Devices may be ordered by using either order number.

National Semiconductor Corporation
2900 Semiconductor Drive
P.O. Box 58090
Santa Clara, CA 95052-8090
Tel: (408) 721-5000
TWX: (910) 339-9240

National Semiconductor GmbH
Westendstrasse 193-195
D-8000 Munchen 21
West Germany
Tel: (089) 5 70 95 01
Telex: 522772

NS Japan Ltd.
Sanseido Bldg. 5F
4-15 Nishi Shinjuku
Shinjuku-Ku,
Tokyo 160, Japan
Tel: 3-299-7001
FAX: 3-299-7000

National Semiconductor Hong Kong Ltd.
Southeast Asia Marketing
Austin Tower, 4th Floor
22-26A Austin Avenue
Tsimshatsui, Kowloon, H.K.
Tel: 3-7231290, 3-7243645
Cable: NSSEAMKTG
Telex: 52996 NSSEA HX

National Semicondutores Do Brasil Ltda.
Av. Brig. Faria Lima, 830
8 Andar
01452 Sao Paulo, SP, Brasil
Tel: (55/11) 212-5066
Telex: 391-1131931 NSBR BR

National Semiconductor (Australia) PTY. Ltd.
21/3 High Street
Bayswater, Victoria 3153
Australia
Tel: (03) 729-6333
Telex: AA32096

National does not assume any responsibility for use of any circuitry described, no circuit patent licenses are implied and National reserves the right at any time without notice to change said circuitry and specifications.

Absolute Maximum Ratings (Note 1)

If Military/Aerospace specified devices are required, please contact the National Semiconductor Sales Office/Distributors for availability and specifications.

Power Supply Voltage	
V_{CC}	+18 V_{DC}
V_{EE}	−18 V_{DC}
Digital Input Voltage, V5–V12	−10 V_{DC} to +18 V_{DC}
Applied Output Voltage, V_O	−11 V_{DC} to +18 V_{DC}
Reference Current, I_{14}	5 mA
Reference Amplifier Inputs, V14, V15	V_{CC}, V_{EE}
Power Dissipation (Note 3)	1000 mW
ESD Susceptibility (Note 4)	TBD

Storage Temperature Range	−65°C to +150°C
Lead Temp. (Soldering, 10 seconds)	
Dual-In-Line Package (Plastic)	260°C
Dual-In-Line Package (Ceramic)	300°C
Surface Mount Package	
Vapor Phase (60 seconds)	215°C
Infrared (15 seconds)	220°C

Operating Ratings

Temperature Range	$T_{MIN} \le T_A \le T_{MAX}$
DAC0808L	−55°C ≤ T_A ≤ +125°C
DAC0808LC Series	0 ≤ T_A ≤ +75°C

Electrical Characteristics

(V_{CC} = 5V, V_{EE} = −15 V_{DC}, V_{REF}/R14 = 2 mA, DAC0808: T_A = −55°C to +125°C, DAC0808C, DAC0807C, DAC0806C, T_A = 0°C to +75°C, and all digital inputs at high logic level unless otherwise noted.)

Symbol	Parameter	Conditions	Min	Typ	Max	Units
E_r	Relative Accuracy (Error Relative to Full Scale I_O)	(Figure 4)				%
	DAC0808L (LM1508-8), DAC0808LC (LM1408-8)				±0.19	%
	DAC0807LC (LM1408-7), (Note 5)				±0.39	%
	DAC0806LC (LM1408-6), (Note 5)				±0.78	%
	Settling Time to Within ½ LSB (Includes t_{PLH})	T_A = 25°C (Note 6), (Figure 5)		150		ns
t_{PLH}, t_{PHL}	Propagation Delay Time	T_A = 25°C, (Figure 5)		30	100	ns
TCI_O	Output Full Scale Current Drift			±20		ppm/°C
MSB V_{IH} V_{IL}	Digital Input Logic Levels High Level, Logic "1" Low Level, Logic "0"	(Figure 3)	2		0.8	V_{DC} V_{DC}
MSB	Digital Input Current High Level Low Level	(Figure 3) V_{IH} = 5V V_{IL} = 0.8V		0 −0.003	0.040 −0.8	mA mA
I_{15}	Reference Input Bias Current	(Figure 3)		−1	−3	μA
	Output Current Range	(Figure 3) V_{EE} = −5V V_{EE} = −15V, T_A = 25°C	0 0	2.0 2.0	2.1 4.2	mA mA
I_O	Output Current	V_{REF} = 2.000V, R14 = 1000Ω, (Figure 3)	1.9	1.99	2.1	mA
	Output Current, All Bits Low	(Figure 3)		0	4	μA
	Output Voltage Compliance (Note 2) V_{EE} = −5V, I_{REF} = 1 mA V_{EE} Below −10V	E_r ≤ 0.19%, T_A = 25°C			−0.55, +0.4 −5.0, +0.4	V_{DC} V_{DC}

National Semiconductor Corporation	National Semiconductor GmbH	NS Japan Ltd.	National Semiconductor Hong Kong Ltd.	National Semiconductor Do Brasil Ltda.	National Semicondutores (Australia) PTY. Ltd.
2900 Semiconductor Drive P.O. Box 58090 Santa Clara, CA 95052-8090 Tel: (408) 721-5000 TWX: (910) 339-9240	Westendstrasse 193-195 D-8000 Munchen 21 West Germany Tel: (089) 5 70 95 01 Telex: 522772	Sanseido Bldg. 5F 4-15 Nishi Shinjuku Shinjuku-Ku, Tokyo 160, Japan Tel: 3-299-7001 FAX: 3-299-7000	Southeast Asia Marketing Austin Tower, 4th Floor 22-26A Austin Avenue Tsimshatsui, Kowloon, H.K. Tel: 3-7231290, 3-7243645 Cable: NSSEAMKTG Telex: 52996 NSSEA HX	Av. Brig. Faria Lima, 830 8 Andar 01452 Sao Paulo, SP. Brasil Tel: (55/11) 212-5066 Telex: 391-1131931 NSBR BR	21/3 High Street Bayswater, Victoria 3153 Australia Tel: (03) 729-6333 Telex: AA32096

National does not assume any responsibility for use of any circuitry described, no circuit patent licenses are implied and National reserves the right at any time without notice to change said circuitry and specifications.

190

 **MOTOROLA**

MC3417, MC3517
MC3418, MC3518

Specifications and Applications Information

CONTINUOUSLY VARIABLE SLOPE DELTA MODULATOR/DEMODULATOR

Providing a simplified approach to digital speech encoding/decoding, the MC3517/18 series of CVSDs is designed for military secure communication and commercial telephone applications. A single IC provides both encoding and decoding functions.

- Encode and Decode Functions on the Same Chip with a Digital Input for Selection
- Utilization of Compatible I^2L — Linear Bipolar Technology
- CMOS Compatible Digital Output
- Digital Input Threshold Selectable ($V_{CC}/2$ reference provided on chip)
- MC3417/MC3517 has a 3-Bit Algorithm (General Communications)
- MC3418/MC3518 has a 4-Bit Algorithm (Commercial Telephone)

CONTINUOUSLY VARIABLE SLOPE DELTA MODULATOR/DEMODULATOR

LASER-TRIMMED INTEGRATED CIRCUIT

L SUFFIX
CERAMIC PACKAGE
CASE 620

PIN CONNECTIONS

Pin	Left		Right	Pin
1	Analog Input	(−)	V_{CC}	16
2	Analog Feedback	(+)	Encode/$\overline{Decode}$	15
3	Syllabic Filter		Clock	14
4	Gain Control		Digital Data Input (−)	13
5	Ref Input (+)		Digital Threshold	12
6	Filter Input (−)		$\overline{Coincidence}$ Output	11
7	Analog Output		$V_{CC}/2$ Output	10
8	V_{EE}		Digital Output	9

CVSD BLOCK DIAGRAM

Encode/Decode 15
Clock 14
Analog Input 1
Analog Feedback 2
Digital Data Input 13
Digital Threshold 12
V_{TH}
Dual Input Comparator
3- or 4-Bit Shift Register
Logic
Coincidence Output 11
Digital Output 9
$V_{CC}/2$ Output 10
$V_{CC}/2$ Ref
I_{Ref}
I_o
Integrator Amplifier
Slope Polarity Switch
V/I Converter
Syllabic Filter 3
Gain Control 4
I_{GC}
I_{Int}
Analog Output (+) 7
Ref Input (+) 5
Filter Input (−) 6

ORDERING INFORMATION

Device	Package	Temperature Range
MC3417L	Ceramic DIP	$0^{\circ}C$ to $+70^{\circ}C$
MC3418L	Ceramic DIP	$0^{\circ}C$ to $+70^{\circ}C$
MC3517L	Ceramic DIP	$-55^{\circ}C$ to $+125^{\circ}C$
MC3518L	Ceramic DIP	$-55^{\circ}C$ to $+125^{\circ}C$

MC3417, MC3517, MC3418, MC3518

MAXIMUM RATINGS
(All voltages referenced to V_{EE}, T_A = 25°C unless otherwise noted.)

Rating	Symbol	Value	Unit
Power Supply Voltage	V_{CC}	−0.4 to +18	Vdc
Differential Analog Input Voltage	V_{ID}	±5.0	Vdc
Digital Threshold Voltage	V_{TH}	−0.4 to V_{CC}	Vdc
Logic Input Voltage (Clock, Digital Data, Encode/$\overline{Decode}$)	V_{Logic}	−0.4 to +18	Vdc
Coincidence Output Voltage	$V_{O(Con)}$	−0.4 to +18	Vdc
Syllabic Filter Input Voltage	$V_{I(Syl)}$	−0.4 to V_{CC}	Vdc
Gain Control Input Voltage	$V_{I(GC)}$	−0.4 to V_{CC}	Vdc
Reference Input Voltage	$V_{I(Ref)}$	$V_{CC}/2$ − 1.0 to V_{CC}	Vdc
$V_{CC}/2$ Output Current	I_{Ref}	−25	mA

ELECTRICAL CHARACTERISTICS
(V_{CC} = 12 V, V_{EE} = Gnd, T_A = 0°C to +70°C for MC3417/18, T_A = −55°C to +125°C for MC3517/18 unless otherwise noted.)

Characteristic	Symbol	MC3417/MC3517			MC3418/MC3518			Unit
		Min	Typ	Max	Min	Typ	Max	
Power Supply Voltage Range (Figure 1)	V_{CCR}	4.75	12	16.5	4.75	12	16.5	Vdc
Power Supply Current (Figure 1) (Idle Channel)	I_{CC}							mA
(V_{CC} = 5.0 V)		−	3.7	5.0	−	3.7	5.0	
(V_{CC} = 15 V)		−	6.0	10	−	6.0	10	
Clock Rate	SR	−	16 k	−		32 k	−	Samples/s
Gain Control Current Range (Figure 2)	I_{GCR}	0.001	−	3.0	0.001	−	3.0	mA
Analog Comparator Input Range (Pins 1 and 2) (4.75 V ≤ V_{CC} ≤ 16.5 V)	V_I	1.3	−	V_{CC} − 1.3	1.3	−	V_{CC} − 1.3	Vdc
Analog Output Range (Pin 7) (4.75 V ≤ V_{CC} ≤ 16.5 V, I_O = ± 5.0 mA)	V_O	1.3	−	V_{CC} − 1.3	1.3	−	V_{CC} − 1.3	Vdc
Input Bias Currents (Figure 3) (Comparator in Active Region)	I_{IB}							μA
Analog Input (I1)		−	0.5	1.5	−	0.25	1.0	
Analog Feedback (I2)		−	0.5	1.5	−	0.25	1.0	
Syllabic Filter Input (I3)		−	0.06	0.5	−	0.06	0.3	
Reference Input (I5)		−	−0.06	−0.5	−	−0.06	−0.3	
Input Offset Current (Comparator in Active Region)	I_{IO}							μA
Analog Input/Analog Feedback |I1−I2| − Figure 3		−	0.15	0.6	−	0.05	0.4	
Integrator Amplifier |I5−I6| − Figure 4		−	0.02	0.2	−	0.01	0.1	
Input Offset Voltage V/I Converter (Pins 3 and 4) − Figure 5	V_{IO}	−	2.0	6.0	−	2.0	6.0	mV
Transconductance	gm							mA/mV
V/I Converter, 0 to 3.0 mA		0.1	0.3	−	0.1	0.3	−	
Integrator Amplifier, 0 to ± 5.0 mA Load		1.0	10	−	1.0	10	−	
Propagation Delay Times (Note 1)								μs
Clock Trigger to Digital Output	t_{PLH}	−	1.0	2.5	−	1.0	2.5	
(C_L = 25 pF to Gnd)	t_{PHL}	−	0.8	2.5	−	0.8	2.5	
Clock Trigger to Coincidence Output	t_{PLH}	−	1.0	3.0	−	1.0	3.0	
(C_L = 25 pF to Gnd) (R_L = 4 kΩ to V_{CC})	t_{PHL}	−	0.8	2.0	−	0.8	2.0	
Coincidence Output Voltage − Low Logic State ($I_{OL(Con)}$ = 3.0 mA)	$V_{OL(Con)}$	−	0.12	0.25	−	0.12	0.25	Vdc
Coincidence Output Leakage Current − High Logic State (V_{OH} = 15.0 V, 0°C ≤ T_A ≤ 70°C)	$I_{OH(Con)}$	−	0.01	0.5	−	0.01	0.5	μA

NOTE 1. All propagation delay times measured 50% to 50% from the negative going (from V_{CC} to +0.4 V) edge of the clock.

ELECTRICAL CHARACTERISTICS (continued)

Characteristic	Symbol	MC3417/MC3517			MC3418/MC3518			Unit
		Min	Typ	Max	Min	Typ	Max	
Applied Digital Threshold Voltage Range (Pin 12)	V_{TH}	+1.2	–	$V_{CC} - 2.0$	+1.2	–	$V_{CC} - 2.0$	Vdc
Digital Threshold Input Current ($1.2\ V \le V_{th} \le V_{CC} - 2.0\ V$)	$I_{I(th)}$							μA
(V_{IL} applied to Pins 13, 14 and 15)		–	–	5.0	–	–	5.0	
(V_{IH} applied to Pins 13, 14 and 15)		–	–10	–50	–	–10	–50	
Maximum Integrator Amplifier Output Current	I_O	±5.0	–	–	±5.0	–	–	mA
$V_{CC}/2$ Generator Maximum Output Current (Source only)	I_{Ref}	+10	–	–	+10	–	–	mA
$V_{CC}/2$ Generator Output Impedance (0 to +10 mA)	z_{Ref}	–	3.0	6.0	–	3.0	6.0	Ω
$V_{CC}/2$ Generator Tolerance ($4.75\ V \le V_{CC} \le 16.5\ V$)	ϵr	–	–	±3.5	–	–	±3.5	%
Logic Input Voltage (Pins 13, 14 and 15)								Vdc
Low Logic State	V_{IL}	Gnd	–	$V_{th} - 0.4$	Gnd	–	$V_{th} - 0.4$	
High Logic State	V_{IH}	$V_{th} + 0.4$	–	18.0	$V_{th} + 0.4$	–	18.0	
Dynamic Total Loop Offset Voltage (Note 2) — Figures 3, 4 and 5	ΣV_{offset}							mV
$I_{GC} = 12.0\ \mu A, V_{CC} = 12\ V$								
$T_A = 25°C$		–	–	–	–	± 0.5	± 1.5	
$0°C \le T_A \le +70°C$ MC3417/18		–	–	–	–	± 0.75	± 2.3	
$-55°C \le T_A \le +125°C$ MC3517/18		–	–	–	–	± 1.5	± 4.0	
$I_{GC} = 33.0\ \mu A, V_{CC} = 12\ V$								
$T_A = 25°C$		–	± 2.5	± 5.0	–	–	–	
$0°C \le T_A \le +70°C$ MC3417/18		–	± 3.0	± 7.5	–	–	–	
$-55°C \le T_A \le +125°C$ MC3517/18		–	±4.5	± 10	–	–	–	
$I_{GC} = 12.0\ \mu A, V_{CC} = 5.0\ V$								
$T_A = 25°C$		–	–	–	–	± 1.0	± 2.0	
$0°C \le T_A \le +70°C$ MC3417/18		–	–	–	–	± 1.3	± 2.8	
$-55°C \le T_A \le +125°C$ MC3517/18		–	–	–	–	± 2.5	± 5.0	
$I_{GC} = 33.0\ \mu A, V_{CC} = 5.0\ V$								
$T_A = 25°C$		–	± 4.0	± 6.0	–	–	–	
$0°C \le T_A \le +70°C$ MC3417/18		–	± 4.5	± 8.0	–	–	–	
$-55°C \le T_A \le +125°C$ MC3517/18		–	± 5.5	± 10	–	–	–	
Digital Output Voltage								Vdc
($I_{OL} = 3.6\ mA$)	V_{OL}	–	0.1	0.4	–	0.1	0.4	
($I_{OH} = -0.35\ mA$)	V_{OH}	$V_{CC} - 1.0$	$V_{CC} - 0.2$	–	$V_{CC} - 1.0$	$V_{CC} - 0.2$	–	
Syllabic Filter Applied Voltage (Pin 3) (Figure 2)	$V_{I(Syl)}$	+3.2	–	V_{CC}	+3.2	–	V_{CC}	Vdc
Integrating Current (Figure 2)	$I_{I(Int)}$							
($I_{GC} = 12.0\ \mu A$)		8.0	10	12	8.0	10	12	μA
($I_{GC} = 1.5\ mA$)		1.45	1.50	1.55	1.45	1.50	1.55	mA
($I_{GC} = 3.0\ mA$)		2.75	3.0	3.25	2.75	3.0	3.25	mA
Dynamic Integrating Current Match ($I_{GC} = 1.5\ mA$) Figure 6	$V_{O(Ave)}$	–	± 100	± 250	–	± 100	± 250	mV
Input Current — High Logic State ($V_{IH} = 18\ V$)	I_{IH}							μA
Digital Data Input		–	–	+5.0	–	–	+5.0	
Clock Input		–	–	+5.0	–	–	+5.0	
Encode/Decode Input		–	–	+5.0	–	–	+5.0	
Input Current — Low Logic State ($V_{IL} = 0\ V$)	I_{IL}							μA
Digital Data Input		–	–	–10	–	–	–10	
Clock Input		–	–	–360	–	–	–360	
Encode/Decode Input		–	–	–36	–	–	–36	
Clock Input, $V_{IL} = 0.4\ V$		–	–	–72	–	–	–72	

NOTE 2. Dynamic total loop offset (ΣV_{offset}) equals V_{IO} (comparator) (Figure 3) minus V_{IOX} (Figure 5). The input offset voltages of the analog comparator and of the integrator amplifier include the effects of input offset current through the input resistors. The slope polarity switch current mismatch appears as an average voltage across the 10 k integrator resistor. For the MC3417/MC3517, the clock frequency is 16.0 kHz. For the MC3418/MC3518, the clock frequency is 32.0 kHz. Idle channel performance is guaranteed if this dynamic total loop offset is less than one-half of the change in integrator output voltage during one clock cycle (ramp step size). Laser trimming is used to insure good idle channel performance.

DEFINITIONS AND FUNCTION OF PINS

Pin 1 — Analog Input

This is the analog comparator inverting input where the voice signal is applied. It may be ac or dc coupled depending on the application. If the voice signal is to be level shifted to the internal reference voltage, then a bias resistor between pins 1 and 10 is used. The resistor is used to establish the reference as the new dc average of the ac coupled signal. The analog comparator was designed for low hysteresis (typically less than 0.1 mV) and high gain (typically 70 dB).

Pin 2 — Analog Feedback

This is the non-inverting input to the analog signal comparator within the IC. In an encoder application it should be connected to the analog output of the encoder circuit. This may be pin 7 or a low pass filter output connected to pin 7. In a decode circuit pin 2 is not used and may be tied to $V_{CC}/2$ on pin 10, ground or left open.

The analog input comparator has bias currents of 1.5 μA max, thus the driving impedances of pins 1 and 2 should be equal to avoid disturbing the idle channel characteristics of the encoder.

Pin 3 — Syllabic Filter

This is the point at which the syllabic filter voltage is returned to the IC in order to control the integrator step size. It is an NPN input to an op amp. The syllabic filter consists of an RC network between pins 11 and 3. Typical time constant values of 6 ms to 50 ms are used in voice codecs.

Pin 4 — Gain Control Input

The syllabic filter voltage appears across C_S of the syllabic filter and is the voltage between V_{CC} and pin 3. The active voltage to current (V–I) converter drives pin 4 to the same voltage at a slew rate of typically 0.5 V/μs. Thus the current injected into pin 4 (I_{GC}) is the syllabic filter voltage divided by the R_x resistance. Figure 6 shows the relationship between I_{GC} (x-axis) and the integrating current, I_{Int} (y-axis). The discrepancy, which is most significant at very low currents, is due to circuitry within the slope polarity switch which enables trimming to a low total loop offset. The R_x resistor is then varied to adjust the loop gain of the codec, but should be no larger than 5.0 kΩ to maintain stability.

Pin 5 — Reference Input

This pin is the non-inverting input of the integrator amplifier. It is used to reference the dc level of the output signal. In an encoder circuit it must reference the same voltage as pin 1 and is tied to pin 10.

Pin 6 — Filter Input

This inverting op amp input is used to connect the integrator external components. The integrating current

(I_{Int}) flows into pin 6 when the analog input (pin 1) is high with respect to the analog feedback (pin 2) in the encode mode or when the digital data input (pin 13) is high in the decode mode. For the opposite states, I_{Int} flows out of Pin 6. Single integration systems require a capacitor and resistor between pins 6 and 7. Multipole configurations will have different circuitry. The resistance between pins 6 and 7 should always be between 8 kΩ and 13 kΩ to maintain good idle channel characteristics.

Pin 7 — Analog Output

This is the integrator op amp output. It is capable of driving a 600-ohm load referenced to $V_{CC}/2$ to +6 dBm and can otherwise be treated as an op amp output. Pins 5, 6, and 7 provide full access to the integrator op amp for designing integration filter networks. The slew rate of the internally compensated integrator op amp is typically 0.5 V/μs. Pin 7 output is current limited for both polarities of current flow at typically 30 mA.

Pin 8 — V_{EE}

The circuit is designed to work in either single or dual power supply applications. Pin 8 is always connected to the most negative supply.

Pin 9 — Digital Output

The digital output provides the results of the delta modulator's conversion. It swings between V_{CC} and V_{EE} and is CMOS or TTL compatible. Pin 9 is inverting with respect to pin 1 and non-inverting with respect to pin 2. It is clocked on the falling edge of pin 14. The typical 10% to 90% rise and fall times are 250 ns and 50 ns respectively for V_{CC} = 12 V and C_L = 25 pF to ground.

Pin 10 — $V_{CC}/2$ Output

An internal low impedance mid-supply reference is provided for use of the MC3417/18 in single supply applications. The internal regulator is a current source and must be loaded with a resistor to insure its sinking capability. If a +6 dBmo signal is expected across a 600 ohm input bias resistor, then pin 10 must sink 2.2 V/600 Ω = 3.66 mA. This is only possible if pin 10 sources 3.66 mA into a resistor normally and will source only the difference under peak load. The reference load resistor is chosen accordingly. A 0.1 μF bypass capacitor from pin 10 to V_{EE} is also recommended. The $V_{CC}/2$ reference is capable of sourcing 10 mA and can be used as a reference elsewhere in the system circuitry.

Pin 11 — Coincidence Output

The duty cycle of this pin is proportional to the voltage across C_S. The coincidence output will be low whenever the content of the internal shift register is all 1s or all 0s. In the MC3417 the register is 3 bits long

DEFINITIONS AND FUNCTIONS OF PINS (continued)

while the MC3418 contains a 4 bit register. Pin 11 is an open collector of an NPN device and requires a pull-up resistor. If the syllabic filter is to have equal charge and discharge time constants, the value of R_P should be much less than R_S. In systems requiring different charge and discharge constants, the charging constant is R_SC_S while the decaying constant is $(R_S + R_P)C_S$. Thus longer decays are easily achievable. The NPN device should not be required to sink more than 3 mA in any configuration. The typical 10% to 90% rise and fall times are 200 ns and 100 ns respectively for R_L = 4 kΩ to +12 V and C_L = 25 pF to ground.

Pin 12 — Digital Threshold

This input sets the switching threshold for pins 13, 14, and 15. It is intended to aid in interfacing different logic families without external parts. Often it is connected to the V_{CC}/2 reference for CMOS interface or can be biased two diode drops above V_{EE} for TTL interface.

Pin 13 — Digital Data Input

In a decode application, the digital data stream is applied to pin 13. In an encoder it may be unused or may be used to transmit signaling message under the control of pin 15. It is an inverting input with respect to pin 9. When pins 9 and 13 are connected, a toggle flip-flop is formed and a forced idle channel pattern

can be transmitted. The digital data input level should be maintained for 0.5 μs before and after the clock trigger for proper clocking.

Pin 14 — Clock Input

The clock input determines the data rate of the codec circuit. A 32K bit rate requires a 32 kHz clock. The switching threshold of the clock input is set by pin 12. The shift register circuit toggles on the falling edge of the clock input. The minimum width for a positive-going pulse on the clock input is 300 ns, whereas for a negative-going pulse, it is 900 ns.

Pin 15 — Encode/$\overline{Decode}$

This pin controls the connection of the analog input comparator and the digital input comparator to the internal shift register. If high, the result of the analog comparison will be clocked into the register on the falling edge at pin 14. If low, the digital input state will be entered. This allows use of the IC as an encoder/decoder or simplex codec without external parts. Furthermore, it allows non-voice patterns to be forced onto the transmission line through pin 13 in an encoder.

Pin 16 — V_{CC}

The power supply range is from 4.75 to 16.5 volts between pin V_{CC} and V_{EE}.

FIGURE 1 — POWER SUPPLY CURRENT

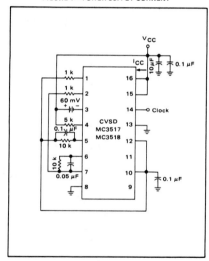

FIGURE 2 — I_{GCR}, GAIN CONTROL RANGE and I_{Int} — INTEGRATING CURRENT

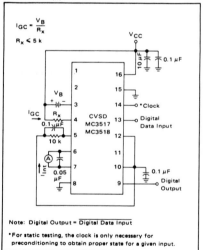

Note: Digital Output = $\overline{\text{Digital Data Input}}$

*For static testing, the clock is only necessary for preconditioning to obtain proper state for a given input.

195

MC3417, MC3517, MC3418, MC3518

TYPICAL PERFORMANCE CURVES

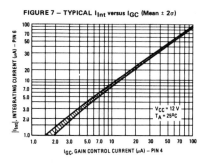

FIGURE 7 – TYPICAL I_{Int} versus I_{GC} (Mean ± 2σ)

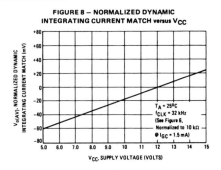

FIGURE 8 – NORMALIZED DYNAMIC
INTEGRATING CURRENT MATCH versus V_{CC}

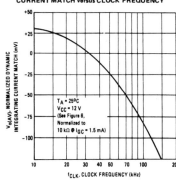

FIGURE 9 – NORMALIZED DYNAMIC INTEGRATING
CURRENT MATCH versus CLOCK FREQUENCY

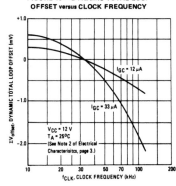

FIGURE 10 – DYNAMIC TOTAL LOOP
OFFSET versus CLOCK FREQUENCY

FIGURE 11 – BLOCK DIAGRAM OF THE CVSD ENCODER

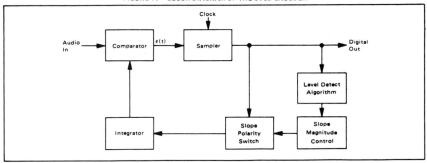

196

MC3417, MC3517, MC3418, MC3518

FIGURE 12 – CVSD WAVEFORMS

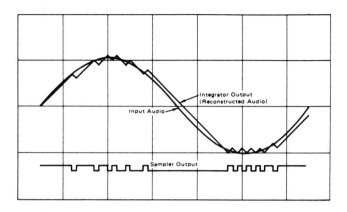

FIGURE 13 – BLOCK DIAGRAM OF THE CVSD DECODER

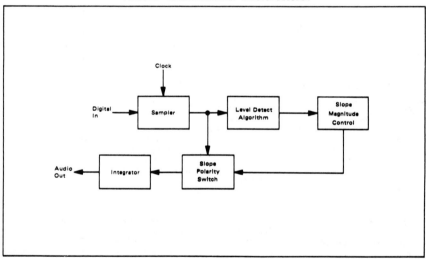

FIGURE 14 — 16 kHz SIMPLEX VOICE CODEC
(Using MC3417, Single Pole Companding and Single Integration)

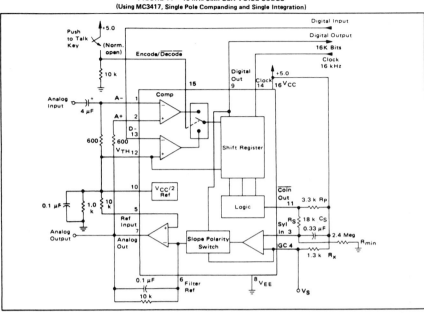

CIRCUIT DESCRIPTION

The continuously variable slope delta modulator (CVSD) is a simple alternative to more complex conventional conversion techniques in systems requiring digital communication of analog signals. The human voice is analog, but digital transmission of any signal over great distance is attractive. Signal/noise ratios do not vary with distance in digital transmission and multiplexing, switching and repeating hardware is more economical and easier to design. However, instrumentation A to D converters do not meet the communications requirements. The CVSD A to D is well suited to the requirements of digital communications and is an economically efficient means of digitizing analog inputs for transmission.

The Delta Modulator

The innermost control loop of a CVSD converter is a simple delta modulator. A block diagram CVSD Encoder is shown in Figure 11. A delta modulator consists of a comparator in the forward path and an integrator in the feedback path of a simple control loop. The inputs to the comparator are the input analog signal and the integrator output. The comparator output reflects the

sign of the difference between the input voltage and the integrator output. That sign bit is the digital output and also controls the direction of ramp in the integrator. The comparator is normally clocked so as to produce a synchronous and band limited digital bit stream.

If the clocked serial bit stream is transmitted, received, and delivered to a similar integrator at a remote point, the remote integrator output is a copy of the transmitting control loop integrator output. To the extent that the integrator at the transmitting locations tracks the input signal, the remote receiver reproduces the input signal. Low pass filtering at the receiver output will eliminate most of the quantizing noise, if the clock rate of the bit stream is an octave or more above the bandwidth of the input signal. Voice bandwidth is 4 kHz and clock rates from 8 k and up are possible. Thus the delta modulator digitizes and transmits the analog input to a remote receiver. The serial, unframed nature of the data is ideal for communications networks. With no input at the transmitter, a continuous one zero alternation is transmitted. If the two integrators are made leaky, then during any loss of contact the receiver output decays to

MC3417, MC3517, MC3418, MC3518

CIRCUIT DESCRIPTION (continued)

zero and receive restart begins without framing when the receiver reacquires. Similarly a delta modulator is tolerant of sporadic bit errors. Figure 12 shows the delta modulator waveforms while Figure 13 shows the corresponding CVSD decoder block diagram.

The Companding Algorithm

The fundamental advantages of the delta modulator are its simplicity and the serial format of its output. Its limitations are its ability to accurately convert the input within a limited digital bit rate. The analog input must be band limited and amplitude limited. The frequency limitations are governed by the nyquist rate while the amplitude capabilities are set by the gain of the integrator.

The frequency limits are bounded on the upper end; that is, for any input bandwidth there exists a clock frequency larger than that bandwidth which will transmit the signal with a specific noise level. However, the amplitude limits are bounded on both upper and lower ends. For a signal level, one specific gain will achieve an optimum noise level. Unfortunately, the basic delta modulator has a small dynamic range over which the noise level is constant.

The continuously variable slope circuitry provides increased dynamic range by adjusting the gain of the integrator. For a given clock frequency and input bandwidth the additional circuitry increases the delta modulator's dynamic range. External to the basic delta modulator is an algorithm which monitors the past few outputs of the delta modulator in a simple shift register. The register is 3 or 4 bits long depending on the application. The accepted CVSD algorithm simply monitors the contents of the shift register and indicates if it contains all 1s or 0s. This condition is called coincidence. When it occurs, it indicates that the gain of the integrator is too small. The coincidence output charges a single pole low pass filter. The voltage output of this syllabic filter controls the integrator gain through a pulse amplitude modulator whose other input is the sign bit or up/down control.

The simplicity of the all ones, all zeros algorithm should not be taken lightly. Many other control algorithms using the shift register have been tried. The key to the accepted algorithm is that it provides a measure of the average power or level of the input signal. Other techniques provide more instantaneous information about the shape of the input curve. The purpose of the algorithm is to control the gain of the integrator and to increase the dynamic range. Thus a measure of the average input level is what is needed.

The algorithm is repeated in the receiver and thus the level data is recovered in the receiver. Because the algorithm only operates on the past serial data, it changes the nature of the bit stream without changing the channel bit rate.

The effect of the algorithm is to compand the input signal. If a CVSD encoder is played into a basic delta modulator, the output of the delta modulator will reflect the shape of the input signal but all of the output will be at an equal level. Thus the algorithm at the output is needed to restore the level variations. The bit stream in the channel is as if it were from a standard delta modulator with a constant level input.

The delta modulator encoder with the CVSD algorithm provides an efficient method for digitizing a voice input in a manner which is especially convenient for digital communciations requirements.

APPLICATIONS INFORMATION
CVSD DESIGN CONSIDERATIONS

A simple CVSD encoder using the MC3417 or MC3418 is shown in Figure 14. These ICs are general purpose CVSD building blocks which allow the system designer to tailor the encoder's transmission characteristics to the application. Thus, the achievable transmission capabilities are constrained by the fundamental limitations of delta modulation and the design of encoder parameters. The performance is not dictated by the internal configuration of the MC3417 and MC3418. There are seven design considerations involved in designing these basic CVSD building blocks into a specific codec application.

These are listed below:
1. Selection of clock rate
2. Required number of shift register bits
3. Selection of loop gain
4. Selection of minimum step size
5. Design of integration filter transfer function
6. Design of syllabic filter transfer function
7. Design of low pass filter at the receiver

The circuit in Figure 14 is the most basic CVSD circuit possible. For many applications in secure radio or other intelligible voice channel requirements, it is entirely sufficient. In this circuit, items 5 and 6 are reduced to their simplest form. The syllabic and integration filters are both single pole networks. The selection of items 1 through 4 govern the codec performance.

TECHNICAL DATA

AN EXCLUSIVE RADIO SHACK SERVICE TO THE EXPERIMENTER

FIBER OPTICS COMMUNICATIONS DATA SYSTEM

MFOE71 - MFOD72 Infrared Emitting Diode and Detector

INTRODUCTION

Explore the frontier of fiber optics data communications with the low cost Archer®
fiber optic emitter — MFOE71 and matching detector MFOD72 pair.

The emitter and detector incorporate integrated connectors which may be
terminated to low-loss 1000 micron plastic fiber optics cable (such as Archer
Cat. No. 276-228.)

ASSEMBLY INSTRUCTIONS

1. Using a single-edge razor blade, strip
the protective fiber jacket (cladding)
back to expose about 1/2 inch of bare
fiber core. Caution: avoid nicking the
fiber core.

2. Scribe the exposed fiber core around
its circumference using the razor blade
so as to leave 1/8 inch of the core
extending beyond the jacket. The
permissable length is 0.100 to 0.180 inch.
Break off the extra core with your fingers.

3. Loosen the cable clamping nut on
each of the respective devices. The nut
may remain mated to the device. Slip
the optical fiber through the clamping
nut connector and guide it into the unit
until the tip of the core seats against
the internal molded lens assembly.
Tighten the connector clamping nut
for a snug fit. Do not overtighten.
The transmission fiber will be locked
into place and ready for use.

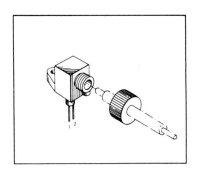

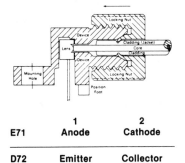

	1	2
E71	Anode	Cathode
D72	Emitter	Collector

CUSTOM PACKAGED IN USA BY RADIO SHACK, A DIVISION OF TANDY CORPORATION

SPECIFICATIONS

MFOE71 Infrared-Emitting Diode

Maximum Ratings

Reverse Voltage	6.0	volts
Forward Current	150.0	mA
Forward Voltage	2.0	volts
Power Dissipation	150.0	mW

Typical Characteristics

Reverse Breakdown Voltage	4.0	volts
Forward Voltage	1.5	volts
Power Launched	165.0	uW
Optical Rise and Fall time	25.0	ns
Peak Wavelength (@ 100mA)	820.0	nm

MFOD72 Infrared-Detector Diode

Maximum Ratings

Collector-Emitter Voltage	30.0	volts
Power Dissipation	150.0	mW
Collector Dark Current	100.0	nA
Saturation Voltage	0.4	volts

Typical Characteristics

Responsivity (@ 5 V)	125	uA/uW
Saturation Voltage	0.25	volts
Turn-on Time	10.0	us
Turn-off Time	60.0	us

RADIO SHACK, A DIVISION OF TANDY CORPORATION

U.S.A.: FORT WORTH, TEXAS 76102
CANADA: BARRIE, ONTARIO L4M 4W5

TANDY CORPORATION

AUSTRALIA	BELGIUM	U.K.
91 KURRAJONG AVENUE MOUNT DRUITT, N.S.W. 2770	RUE DESPIED'S D'ALOUETTE, 39 5140 NANINNE (NAMUR)	BILSTON ROAD WEDNESBURY WEST MIDLANDS WS10 7JN

Printed in U.S.A.